Molecules and Clusters in Intense Laser Fields

This book provides a thorough and comprehensive introduction to the physics of molecules and clusters in intense laser fields. It covers both theoretical and experimental aspects of the subject, and presents new research in the area of clusters in intense laser fields. Topics covered include coherent control, diatomic and polyatomic molecules, and production and diagnostics of femtosecond pulses.

Written by leading researchers in the field, this book will be of interest to graduate students and researchers in atomic, molecular and optical physics. It will also be suitable as a reference text for advanced physics courses.

JAN POSTHUMUS obtained his PhD from the University of Groningen. He then did research in Reading at the Rutherford Appleton Laboratory, publishing numerous papers and journal articles on the subject of molecules in intense laser fields. He is currently a research physicist at the University of Munich.

Molecules and Clusters in
Intense Laser Fields

Edited by Jan Posthumus
University of Munich

CAMBRIDGE
UNIVERSITY PRESS

PUBLISHED BY THE PRESS SYNDICATE OF THE UNIVERSITY OF CAMBRIDGE
The Pitt Building, Trumpington Street, Cambridge, United Kingdom

CAMBRIDGE UNIVERSITY PRESS
The Edinburgh Building, Cambridge CB2 2RU, UK
40 West 20th Street, New York, NY 10011–4211, USA
10 Stamford Road, Oakleigh, VIC 3166, Australia
Ruiz de Alarcón 13, 28014 Madrid, Spain
Dock House, The Waterfront, Cape Town 8001, South Africa

http://www.cambridge.org

First published 2001

Printed in the United Kingdom at the University Press, Cambridge

Typeset by the author [CRC]

A catalogue record for this book is available from the British Library

Library of Congress Cataloguing in Publication data
Posthumus, Jan, 1964–
Molecules and clusters in intense laser fields / Dr Jan Posthumus.
p.   cm.
Includes bibliographical references and index.
ISBN 0 521 77240 0
1. Photoionization.   2. Multiphoton processes.   3. Laser pulses, Ultrashort.   I. Title.
QC702.7.P8P68   2001
539′.6–dc21   00-052934

ISBN 0 521 77240 0 hardback

Contents

Preface

Young researchers, when they have just joined a research group, are all too often sent to the library with a photocopying card and the assurance that all the necessary background information can be found in the journals. Those following this advice soon start showing signs of sleepless nights and attacks of panic. The student has studied physics for a number of years, believing it to be the mother of all sciences, but suddenly existential thoughts make all other vocations seem much more relevant. At this stage the risk of dropping out of science is very high. However, those who succeed along this path are destined for a successful academic career and will proudly give the same advice to the next generation. Fortunately, most students are wise enough to approach a young colleague, who is all too familiar with the problem of where to find relevant information for the real beginner.

This book is aimed primarily at postgraduate students and postdoctoral research assistants. The research area of 'molecules and clusters in intense laser fields' is itself quite young and also rapidly developing. Not surprisingly, a proper introduction has been lacking. It seemed appropriate, therefore, to write an introductory text, particularly since interest in the subject is growing. Since the young researcher would normally approach his peers for practical information, it was thought beneficial for the scope of this book if the authors themselves were quite young. In order to give the book some breadth, the style of an edited volume was chosen. This has made it possible to include the latest advances in all the sub-topics.

The most widely used laser system in intense-laser-field investigations is the Ti:sapphire laser. The advances in 'all-solid-state-technology' femtosecond lasers during the past decade have been tremendous. The first chapter discusses the state of the art of production and diagnostics of femtosecond pulses from the point of view of a laser designer and operator.

The second and third chapters treat the physics as well as the experimental aspects of diatomic and polyatomic molecules in intense laser fields.
Chapter 4 is on the interesting subject of coherent control. A rather
new research area is the interaction of intense lasers with clusters. This
very hot topic (quite literally) has already aroused much interest. The
comprehensive overview of the experimental techniques, the most important results and their interpretation in the last three chapters is therefore
timely.

It has been said that the key to success is a good teacher. This book
cannot replace a good teacher, but it is hoped that it will assist in building
up some elementary understanding of the subject. For those actively
pursuing research, further study of the original articles, which are cited
wherever appropriate, will be essential. The exhaustive reference listings
are located at the end of each chapter.

Finally, I would like to thank all of the co-authors for their great contributions. Sincere thanks are also due to the wonderful people of Cambridge
University Press, who have been very supportive ... and very patient.
On behalf of all the authors, I wish the readers many new insights and
may their enthusiasm for the subject be as great as ours.

Jan Posthumus

1

Ultra-high-intensity lasers based on Ti:sapphire

Philip F. Taday and Andrew J. Langley

Central Laser Facility
CLRC Rutherford Appleton Laboratory
Chilton, Didcot OX11 0QX, UK

1.1 Introduction

Naturally, the scientific study of matter under extreme conditions is of fundamental interest, but it is also true that its direction and progress are largely determined by technological innovations. Also theorists generally stay within the bounds of what is (nearly) feasible, for even the best theories require verification. The technological development underlying the subject of this book is that of high-power lasers. Particularly the advent of affordable, ultra-short-pulsed lasers has considerably intensified the study of molecules and clusters in intense fields.

Two determining technical advances in this respect were the development of chirped pulse amplification (CPA) [1] and the self-mode-locked Ti:sapphire laser [2] (which has generated pulses as short as 7 fs [3]). It is thus possible to build laboratory-scale systems that produce high-energy pulses (3–4 mJ in 20 fs), even at kilohertz repetition rates [4]. An even more powerful system, a 100-TW, sub-20-fs laser system was demonstrated by Yamakawa *et al.* [5], but at the lower repetition rate of 10 Hz. Extremely short pulses of only 4.5 fs – but still with energies as high as 70 μJ – were achieved with pulse compression techniques by Nisoli *et al.* [6].

Particularly energetic laser pulses are required for the study of inertial-confinement fusion (ICF). At the Lawrence Livermore National Laboratory [7], for example, 660 J in a 440 ± 20-fs pulse can be focused down to an intensity $>7 \times 10^{20}$ W cm^{-2}. Meanwhile, researchers at the CLRC Rutherford Appleton Laboratory have achieved intensities greater than 10^{19} W cm^{-2} using the VULCAN laser [8]. Figure 1.1 shows the disc amplifiers of this laser system. With these big lasers, which are limited to a few shots per hour, the amplified spontaneous emission (ASE) generally causes a serious problem, since it is very energetic and interacts with the

Fig. 1.1.   The disc amplifiers on the Vulcan laser system at the CLRC Rutherford Appleton laboratory. Courtesy of the CLRC Rutherford Appleton Laboratory.

target prior to the arrival of the main pulse. Ross *et al.* [9] have therefore
suggested a novel method that combines the techniques of CPA with optical parametric amplification to produce ultra-short, ultra-high-intensity
pulses with a low pre-pulse (OPCPA). The development of OPCPA could
possibly lead to focused intensities as great as 10^{22} W cm^{-2}.

The progress in laser development has opened up new fields of research
in high-field interaction physics. Some of these topics will be covered in
later sections of this book. With ultra-high intensities $>10^{19}$ W cm^{-2},
advances have been made in the acceleration of plasma electrons to over
100 MeV [10], in the observation of relativistic self-focusing and chan-
nelling in low-density plasmas [11] and the activation of the nucleus by
the interaction with γ-rays from a laser-produced plasma [12]. In the
high-intensity regime 10^{15-17} W cm^{-2}, much progress has been made in
understanding the fundamental interactions of atoms and molecules with
intense laser fields [13]. There have also been significant developments in
applied areas, such as the detection of trace compounds [14] and the use
of ballistic photons in medical diagnostics. With the wide bandwidths
available from short pulses, progress in telecommunication with terahertz
information transmission is now becoming possible. The advantage of
using intense short pulses for material processing has also been demonstrated [15].

1.2 System considerations in building a Ti : sapphire system

An oscillator generates ultra-short pulses at relatively low energies; typically a few nanojoules at repetition rates of hundreds of megahertz. Although these pulses can be focused to intensities as great as 10^{13} W cm^{-2} [16], this intensity is still too low for many researchers and amplification of the pulses is required.

Direct amplification of ultra-short pulses is limited to peak intensities of a few GW cm^{-2}, due to non-linear effects. One example of such a non-linearity is the intensity-dependent refractive index of a material,

$$n = n_1 + n_2 I, \tag{1.1}$$

where n_1 is the refractive index, n_2 is the non-linear refractive index of the material and I is the intensity of the laser beam. Non-linear effects lead to wavefront destruction and in extreme cases whole-beam self-focusing into filaments, with the potential of destroying optical components. To avoid these problems, the peak intensity of the amplified beam must be limited to a few GW cm^{-2}. On the other hand, the most efficient extraction of the energy stored in the amplifier is achieved by operating near saturation; for a Ti : sapphire amplifier medium this means at an input fluence of a few J cm^{-2}. However, the intensity ceiling, combined with the short pulselengths of e.g. 100 fs, ordinarily limits the input fluence to a few mJ cm^{-2}.

The solution to this apparent contradiction is a technique known as chirped pulse amplification (CPA) [1]. CPA was first used in radar technology to generate high-energy radar pulses with a short pulse duration. It involves stretching the duration of a low-energy short pulse so as to reduce its peak power during amplification. After amplification, the short duration is recovered by reversing the initial stretching process. In practice, the pulse stretching and subsequent compression processes involve passing the short pulses through dispersive optical systems, usually consisting of diffraction gratings.

The fidelity of the recovered pulse depends greatly on the design of the stretcher and compressor. To get a better understanding of the limitations on the pulse fidelity, let us perform a Taylor expansion of the frequency-dependent phase, $\phi(\omega)$, of the laser pulse:

$$\phi(\omega) = \phi(\omega_0) + \phi'(\omega_0)(\omega - \omega_0) + \frac{1}{2}\phi''(\omega_0)(\omega - \omega_0)^2$$
$$+ \frac{1}{6}\phi'''(\omega_0)(\omega - \omega_0)^3 + \frac{1}{24}\phi''''(\omega_0)(\omega - \omega_0)^4 + \cdots, \tag{1.2}$$

where ω is the angular frequency and ω_0 is the centre frequency of the pulse. The first term in the expansion is a constant, whilst the second term

Table 1.1.   *The magnitude of each phase distortion term that gives rise to a 10% broadening of an input bandwidth-limited Gaussian pulse*

Pulse duration (fs)	GVD ($fs^2 \, rad^{-1}$)	TOD ($fs^3 \, rad^{-2}$)	FOD ($fs^4 \, rad^{-3}$)
10	17.7	68	261
50	440	8450	163 030
100	1770	68 000	2 603 500

represents the time delay of the pulse. In third place, ϕ'', is the second-order dispersion term; it is also often referred to as linear chirp. This important term is responsible for the group-velocity dispersion (GVD) and determines the amount of pulse stretching and compression in a CPA system. The fourth term, ϕ''', is known as the third-order dispersion (TOD) or cubic phase term.

The duration of the dispersed pulse is related to the accumulated phase by [17]

$$\tau(\omega) = \frac{\partial \phi(\omega)}{\partial \omega}. \tag{1.3}$$

Thus, by substituting equation (1.2) into (1.3), we get

$$\tau(\omega) = \phi' + \phi''(\omega - \omega_0) + \frac{1}{2}\phi'''(\omega - \omega_0)^2$$
$$+ \frac{1}{6}\phi''''(\omega - \omega_0)^3 + \text{higher-order terms.} \tag{1.4}$$

For a well-behaved CPA system we assume that the effect of each term of the expansion produces a pulse broadening or distortion that is significantly smaller than the effect of the previous term. Therefore, we can write

$$\phi''(\omega - \omega_0) \gg \frac{1}{2}\phi'''(\omega - \omega_0)^2 \gg \frac{1}{6}\phi''''(\omega - \omega_0)^3 \cdots. \tag{1.5}$$

In table 1.1 we give estimates of the maximum phase contribution allowed to give a 10% broadening of the initial, Gaussian input pulse. The magnitude of each phase term is given in terms of $fs^n \, rad^{n-1}$, where n is the order of the term. Thus, to obtain less than 10% broadening of a 10-fs input pulse, the GVD must be $<17.7 \, fs^2 \, rad^{-1}$. To avoid further significant broadening, the higher-order terms should of course be smaller than the values given in the table.

All CPA systems consist of some form of stretcher, amplifier and compressor. Therefore, when one is considering phase contributions in a full system, it is important to consider not only the contributions due to the stretcher and compressor but also those of all the other materials in the

Table 1.2. *The group velocity, third- and fourth-order dispersion terms, for pulses propagating through 1 cm of common optical materials*

Optical material	GVD (fs^2)	TOD (fs^3)	FOD (fs^4)
BK7	446	321	-106
Sapphire (ordinary axis)	581	442	-155
Sapphire (extraordinary axis)	566	414	-155
KDP	377	249	-56
Calcite	455	331	-115

system, including air. The delay, when a pulse propagates through a medium, is given by [17]

$$\tau(\omega) = \frac{l}{c}\frac{\mathrm{d}}{\mathrm{d}\omega}\left(n\left(\omega\right)\omega\right), \tag{1.6}$$

where ω is the angular frequency, l is the path length in the material, c is the speed of light and $n(\omega)$ is the refractive index of the material. The refractive index can be calculated from the Sellmeier equations, which can be found in optics catalogues. By taking the derivatives of the refractive index with respect to wavelength, it is possible to calculate the GVD, TOD and FOD terms for the material of propagation. The GVD term is given by [17]

$$\frac{\mathrm{d}^2\phi(\omega)}{\mathrm{d}\omega^2} = \frac{\lambda^3 l}{2\pi c^2}\frac{\mathrm{d}^2 n(\lambda)}{\mathrm{d}\lambda^2}, \tag{1.7}$$

where λ is the wavelength of the laser pulse under consideration, l is the length of the material and c is the speed of light. Similarly the TOD and FOD terms are given by [17]

$$\frac{\mathrm{d}^3\phi(\omega)}{\mathrm{d}\omega^3} = \frac{\lambda^4 l}{4\pi^2 c^3}\left(3\frac{\mathrm{d}^2 n(\lambda)}{\mathrm{d}\lambda^2} + \lambda\frac{\mathrm{d}^3 n(\lambda)}{\mathrm{d}\lambda^3}\right), \tag{1.8}$$

$$\frac{\mathrm{d}^4\phi(\omega)}{\mathrm{d}\omega^4} = \frac{\lambda^5 l}{8\pi^3 c^4}\left(12\frac{\mathrm{d}^2 n(\lambda)}{\mathrm{d}\lambda^2} + 8\lambda\frac{\mathrm{d}^3 n(\lambda)}{\mathrm{d}\lambda^3} + \lambda^2\frac{\mathrm{d}^4 n(\lambda)}{\mathrm{d}\lambda^4}\right). \tag{1.9}$$

For the design of a CPA system, depending on the shortness of the pulses we are trying to achieve, it might be necessary to consider only terms up to the third-order dispersion term. To give an idea of the size of the above phase terms, table 1.2 gives the values for some of the more common materials encountered in an amplifier system, for 1 cm of material.

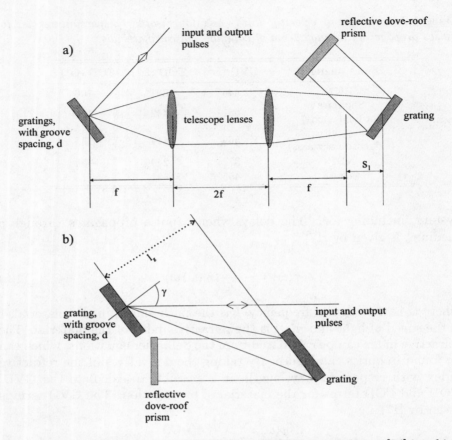

Fig. 1.2.   (a) A schematic diagram of the original stretcher design [24] in which two identical lenses, each of focal length f, are separated by $2f$ and placed between a pair of antiparallel gratings. One of the gratings can be at the focus of one the lenses; the other is placed a distance s_1 from the focus. The effective distance between the gratings is $l_g = 2f - s_1$. (b) A pulse compressor.

We now consider the main sources of dispersion in a CPA system, namely the pulse stretcher and the compressor. The GVD and TOD are given by [18]

$$\frac{\mathrm{d}^2\phi(\omega)}{\mathrm{d}\omega^2} = \frac{\lambda^3 l_g}{2\pi c^2 d^2}\left[1 - \left(\frac{\lambda}{d} - \sin\gamma\right)^2\right]^{-\frac{3}{2}}, \qquad (1.10)$$

$$\frac{\mathrm{d}^3\phi(\omega)}{\mathrm{d}\omega^3} = \frac{3\lambda}{2\pi c}\frac{\mathrm{d}^2\phi(\omega)}{\mathrm{d}\omega^2}\left(\frac{1 + (\lambda/d)\sin\gamma - \sin^2\gamma}{1 - ((\lambda/d) - \sin\gamma)^2}\right). \qquad (1.11)$$

The angle γ and the grating separation l_g are defined in figure 1.2; d is the groove spacing of the gratings. It may be useful here to consider the magnitudes of the GVD and TOD terms in a typical CPA system. For a compressor working at the Littrow angle (given by $\sin\gamma = \frac{1}{2}\lambda/d$) and with 1500 lines mm^{-1}, we have $l_g = 1100$ mm. The GVD and TOD then are -4.38×10^6 fs^2 and 1.187×10^7 fs^3, respectively. These are very large terms and, considering the phase terms, one can easily see that a small mismatch between the compressor and the stretcher gratings in terms of l_g or the angle γ will lead to large errors.

The behaviours of all the optical components discussed so far are well understood, but there are other components within a CPA that are not so well behaved. For example, mirror manufacturers are producing mirrors with very low GVD, but having large TOD terms. In a practical system this makes compensating for TOD harder. In most systems, the compressor-grating incidence is changed to correct for TOD. Since changing the compressor-grating angle of incidence, γ, also alters the GVD by large amounts, iterating the changes in the angle of incidence and the grating separation becomes necessary; this is a tedious and imprecise procedure.

Oscillator design

Prior to 1990, those groups wanting to work with short pulses (we define short as less than 500 fs) had to use synchronously pumped, mode-locked dye lasers. The pulses from these 'difficult' lasers were normally amplified in dye-based systems and the output was generally not transform limited. The pulse energy was limited to a few millijoules, due to the short storage time (10–12 ns) and the low saturation fluence (0.002 J cm^{-2}) of the dye. In principle, these lasers could be tuned by changing the dye in the oscillator cell. In practice, however, this was rather inconvenient. A revolution occurred in oscillator technology in the 1990s with the demonstration, by Sibbett's group at the University of St Andrews, of self-mode-locked Ti : sapphire lasers [2]. The introduction of Ti : sapphire to the short-pulse-laser community has many benefits: a very large gain bandwidth of around 400 nm, a high thermal conductivity, a long storage time of 3 μs and a high saturation fluence of about 1 J cm^{-2}. The last two properties make Ti : sapphire an extremely interesting material for high-energy amplification.

The self-mode-locking mechanism in a Ti : sapphire oscillator relies on a non-linear Kerr lens effect in the laser crystal [2, 18]. Other crystals have been used to generate short pulses, for example, Cr : LiSAF [19] and Yb : silica [20], but we concern ourselves only with Ti : sapphire.

Figure 1.3(a) is a schematic diagram of a self-mode-locked Ti : sapphire laser. The laser cavity is very simple and consists of an end mirror, an output coupler, a pair of prisms, focusing fold mirrors and the crystal itself.

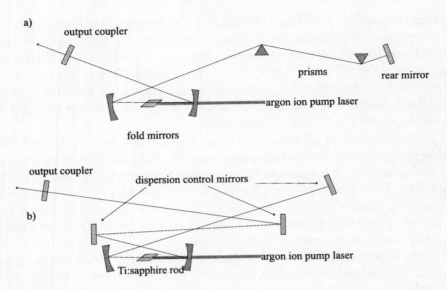

Fig. 1.3.   Diagram of a self-mode-locked Ti : sapphire oscillator with prisms (a) and without prisms (b). Note the angle of the curved fold mirrors, which allows compensation of the astigmatism that is introduced by the Brewster-angle-cut Ti : sapphire rod (see for example [21]).

A continuous-wave pump laser is focused onto the Ti : sapphire crystal. (The large-frame argon-ion lasers have been superseded by frequency-doubled CW diode-pumped Nd : YVO$_4$ lasers.) The Ti : sapphire crystal is normally Brewster-cut with the astigmatism compensated for by a pair of concave mirrors in an off-axis 'Z' or 'X' configuration [21]. The pump laser and the Ti : sapphire beams are collinear within the crystal. The pair of prisms compensates for the dispersion of the Ti : sapphire crystal [22]. Instead, one may use dispersion-compensating mirrors [23] and thus remove all tunable components from the cavity. Whilst this improves the stability, it also fixes the central wavelength. With this method, pulses as short as 5 fs [3] have been generated. A schematic layout is shown in figure 1.3(b).

Stretcher and compressor designs

A short pulse-grating stretcher derives its origin from a scheme suggested by Martinez [24] in which pulses are negatively chirped. The stretcher comprises a telescope, with a magnification of one, between two anti-parallel gratings, as shown in figure 1.2. Pessot [25] showed that the chirp introduced by Martinez' stretcher can be compensated by a pair of similar gratings in a parallel configuration, as demonstrated by Treacy [26] in 1969. The most compact expression for the group delay in a stretcher

Table 1.3.   *Non-linear refractive indices of some common optical materials*

Material	Wavelength (nm)	n_2 (cm^2 W^{-1})	Reference
Air	800	2.9×10^{-19}	[27]
–	600	3.3×10^{-19}	[18]
–	248	12×10^{-19}	[28]
BK7 glass	804	3.75×10^{-16}	[18]
Fused silica	804	3.21×10^{-16}	[29]
–	600	3.0×10^{-16}	[18]
SF10	600	1.3×10^{-15}	[18]
Sapphire	780	8.0×10^{-16}	[18]
Mg$_2$F	804	1.15×10^{-16}	[29]
BBO	1064	5.0×10^{-16}	[30]
–	532	4.8×10^{-16}	[30]
LBO	1064	2.1×10^{-16}	[30]
–	532	2.8×10^{-16}	[30]
KTP	1064	4.6×10^{-15}	[30]
–	532	3.7×10^{-15}	[30]
Water	804	5.7×10^{-16}	[29]
O$_2$	800	5.1×10^{-19}	[27]
–	248	3.0×10^{-18}	[28]
N$_2$	800	2.3×10^{-19}	[27]
–	248	7.5×10^{-19}	[28]
Xe	800	8.1×10^{-19}	[27]
–	248	-9.6×10^{-17}	[28]
Ar	800	1.4×10^{-19}	[27]
SF$_6$	800	1.6×10^{-19}	[27]

was given by Martinez [24]:

$$\phi''(\omega) = \frac{2\omega l_{\mathrm{g}}}{c}\left[1 - \left(\frac{2\pi c}{\omega d} - \sin\gamma\right)^2\right]^{\frac{1}{2}}, \qquad (1.12)$$

where ω is the angular frequency, l_{g} is the distance between the gratings, d is the groove density, c is the speed of light and γ is the angle of incidence. This expression is for a single-pass compressor.

We must consider what stretching factor is required in an amplifier. The approximate value of the non-linear phase contribution is given by

$$B = \frac{2\pi n_2}{\lambda}\int_0^z I(t, z)\,\mathrm{d}z, \qquad (1.13)$$

where n_2 is the non-linear refractive index of a material (typical values are shown in table 1.3) and $I(t, z)$ is the intensity of the laser pulse in the medium under consideration. Equation (1.13) is an expression known as the 'B-integral'; its value should be kept below unity in any laser amplifier. If B exceeds this value, break-up of the beam can occur, with the result that the laser pulse fails to reach its target energy or focused intensity. Consequently, in most high-energy, Ti:sapphire ultra-short-pulse amplifiers, the pulse must be stretched to between 300 ps and 1 ns.

Conventional pulse stretchers normally include a lens telescope system. However, for pulse durations below 100 fs, these refractive optics introduce such large chromatic aberrations in the beam that it is necessary to use all-reflective optics. There are several designs of pulse stretcher currently in use. The complex design by Lemoff and Barty [31] allows a 10-fs pulse to be stretched to about 1 ns. The system introduces some fourth-order phase aberration to the pulse, which compensates for the fourth-order phase errors resulting from the amplifier and this allows compression down to 18 fs. The bandwidth of the stretcher is approximately 130 nm. All angles in this system have to be set to within 1 min and distances between optics to within 250 μm. Groups at Michigan [32] and at the Laboratoire d'Optique Appliquée [33] have both used an Offner [34] telescope inside their stretcher. This design uses only one grating but, due to the high number of reflections occurring in the stretcher, a high tolerance on the optical flatness of the components, $\lambda/20$ or better, is required. The Offner triplet telescope, when it is aligned correctly, should give aberration-free laser beams. Both groups stretch a 30-fs pulse to 300 ps and have demonstrated compression back down to 33 fs [33]. The design by Itatani *et al.* [35] is also based on the Offner telescope but, because this design uses two gratings and thus uses fewer reflections, a lower tolerance on the telescope optics of $\lambda/10$ is required.

In summary, when one is designing and building a stretcher for a short-pulse system, one must consider

1. the stretch required for the total amount of amplification,

2. the transmitted spectral bandwidth of the system,

3. the optical aberrations introduced into the laser beam by running the stretcher in an off-axis geometry and

4. the surface accuracy of the optics.

In order for a laser designer to assess these factors properly, some form of ray-tracing programming is required.

1.3 Laser-pulse amplification

Prior to the mid-1980s and the appearance of Ti:sapphire lasers, amplification of short pulses was done in organic-dye lasers. These systems were

Table 1.4.   *Optical properties of possible materials used in short-pulse amplifiers*

Material	λ_{laser} (nm)	$\delta\lambda$ (nm)	σ (cm^2)	J_{sat} (J cm^{-2})	Storage time (s)	Typical pump
Dyes	300–1000	$\simeq$50	$>10^{-16}$	$\simeq$0.001	$10^{-8}\ldots{-12}$	Laser
KrF	248	$\simeq$2	3×10^{-16}	0.003	$<10^{-8}$	Discharge
Alexandrite	750	$\simeq$100	7×10^{-21}	88.2	2.6×10^{-4}	Flashlamp
Ti : sapphire	790	$\simeq$400	3×10^{-19}	0.88	3×10^{-6}	Laser
Cr : LiSAF	830	$\simeq$250	5×10^{-20}	8	6×10^{-5}	Flashlamp
Nd : glass-phosphate	1053	22	4×10^{-20}		3×10^{-4}	Flashlamp
Nd : glass-silicate	1065	28	2.3×10^{-20}		3×10^{-4}	Flashlamp

limited to a few millijoules because of the low saturation fluences (see table 1.4),

$$J_{\text{sat}} = \frac{h\nu}{\sigma}, \tag{1.14}$$

where ν is the laser transition frequency, h is Planck's constant and σ is the gain cross section. The amplifiers of the 1980s, which used dyes with a saturation fluence of 1 or 2 mJ cm^{-2}, were capable of generating energies of 5 mJ with dye cells as large as 2 cm. With the advent of new solid-state materials, such as Ti : sapphire with $J_{\text{sat}} = 0.88$ J cm^{-2}, energies in excess of 2 J can be generated in crystals of less than 2 cm diameter. Coupled with Ti : sapphire's extremely large gain cross section of more than 400 nm, amplified pulses as short as 20 fs can be obtained.

Most 'table-top' multi-terawatt-amplifier laser systems consist of a high-gain pre-amplifier, followed by several (normally two) power amplifiers. The pre-amplifier is designed to increase the energy output from the nanojoule to the millijoule level with a nett gain in the region of 10^7. With the power amplifiers the pulses can finally reach the joule level.

There are two basic types of pre-amplifier designs, the regenerative amplifiers, known as 'regens', and the multi-pass configurations. The regenerative amplifier, shown schematically in figure 1.4(a), is very similar to a linear laser cavity. A-low energy oscillator pulse can be injected into the cavity via a thin polariser and is trapped by switching the polarisation with a Pockels cell. The pulse travels through the amplifier medium, gaining energy. A regenerative amplifier is run at low gain per pass to suppress the build-up of amplified spontaneous emission (ASE). This means that the pulse must make in the region of 20–25 passes around the cavity before being switched out by the Pockels cell. An example of a regenerative is given in a paper by Yamakawa *et al.* [36].

A regenerative amplifier allows easy alignment of the whole laser system. The cavity-dumped light from the unseeded regenerative amplifier, that is with the oscillator blocked, can be used to align the rest of the laser system. The output beam from the 'regen' should be a good-quality

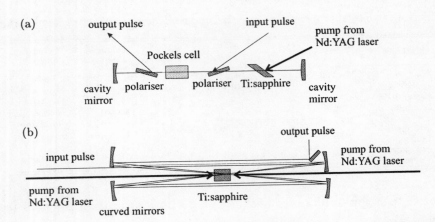

Fig. 1.4. Configurations of (a) a regenerative amplifier and (b) a multi-pass amplifier. Note the angle of the thin-film polarisers in the cavity of the regenerative amplifier. In order to work at the very high bandwidths of >100 nm, these polarisers are designed to work at angles of about 72°.

TEM_{00} mode, which will provide an excellent seed for the subsequent multi-pass amplifier. The major disadvantage of a regenerative amplifier in a short-pulse laser system is the large number of passes and, consequently, the long optical path through high-index materials, which leads to additional difficulty in recovering the short pulse.

Figure 1.4(b) shows a multi-pass configuration in which the amplified beam makes several passes through the gain medium without the use of the cavity arrangement present in a regen. The main advantage of this design is that it can be run at a much higher gain than can an equivalent regenerative system, thus requiring fewer passes through the gain medium. Using this type of amplifier, pulses as short as 20 fs have been amplified to energies of a few hundred microjoules [37].

An example of a complete system

Figure 1.5 shows the layout of the newly built femtosecond terawatt laser at the CLRC Rutherford Appleton Laboratory, UK. It is built on three 1.5×3.0-m² optical tables in a linear configuration. Pulses for amplification are derived from a mirror-dispersion-controlled Ti : sapphire oscillator [38] (Femto, Technische Universität, Vienna), pumped by a frequency-doubled CW diode-pumped $Nd : YVO_4$ laser (Millennia, Spectra-Physics).

The Femto laser provides pulses at a repetition rate of 76 MHz, centred around 790 nm, with sufficient bandwidth for compression to 20 fs. Prior to amplification, the pulses are stretched to 300 ps in a pulse stretcher. The stretcher, shown schematically in figure 1.6, is essentially a simple all-reflective analogue of a dual-grating stretcher with a single relay lens.

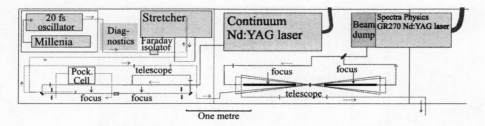

Fig. 1.5. The layout of the newly built femtosecond terawatt laser at the CLRC Rutherford Appleton Laboratory, UK.

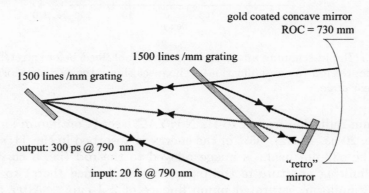

Fig. 1.6. A schematic diagram of the stretcher used on the ASTRA terawatt laser system.

It is designed with a bandpass of four times the FWHM bandwidth of a 30-fs Gaussian pulse.

The input pulses make a double pass through the stretcher which consists of two 1500-lines mm^{-1} gratings (Spectrogon), a large gold coated mirror (ROC = 730 mm) and a gold-coated flat mirror (Optical Surfaces). The gratings and mirrors were specified to $\lambda/10$ and $\lambda/20$ flatnesses, respectively. For this design, minimum beam aberration is achieved by retro-reflecting the beam. It was found necessary to place a Faraday isolator before the stretcher to minimise the likelihood of back reflections disrupting the oscillator.

The stretched pulses are amplified in the first amplifier to an energy of up to 2 mJ. This amplifier consists of four confocal mirrors (ROC = 2000 mm), which direct and focus the amplified beam through a 7-mm long Ti : sapphire rod (Crystal Systems, FOM = 150 and $\alpha_{514} = 4.7$). The parameter α_λ denotes the absorption coefficient at wavelength λ. The FOM or 'figure of merit' is the ratio of α_{514} and α_{820}, i.e. the pump and emission wavelengths. It is influenced by the concentration of Ti^{3+} and Ti^{4+} ions in the crystal [39]. The rod is pumped with 140-mJ, 20-ns pulses

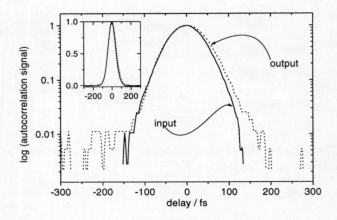

Fig. 1.7. Rapid-scanning autocorrelator traces obtained before stretching and after compression (no amplification). The inset shows the two autocorrelations on a linear scale.

of 532-nm radiation from a 10-Hz Nd : YAG laser (Continuum long-pulse Surelite, 20 ns, 10 Hz). 95% of the energy is absorbed in the Ti : sapphire rod. The 532-nm beam is image relayed to the rod (i.e. a clear image of the limiting aperture in the pump laser is formed there) to provide a nearly uniform, saturated pump fluence of 2 J cm^{-2}. After the first five passes, the pre-amplified pulse train is extracted from the amplifier and passed through a Pockels cell (Leysop) which allows a 'pulse pick' from the 76-MHz pulse train and also minimises the amount of amplified fluorescence in the high-gain first amplifier. The selected pulse is then re-injected into the amplifier to boost its energy to the 1–2 mJ level. The amplifier provides a nett gain of 4×10^6.

At this stage the pulses can be compressed to 50 fs, using a conventional compressor design comprised of a pair of 1500-lines mm^{-1} gratings (Spectrogon) in a parallel arrangement. The transmission efficiency of the compressor is 50%. During this phase of the development programme an older type Ti : sapphire oscillator (Spectra-Physics, Tsunami) was used as the source of 50-fs pulses for amplification. The stretcher and compressor were set up using rapid-scanning autocorrelators with unamplified oscillator pulses to test that the stretched pulses were successfully recompressed to 50-fs duration. Figure 1.7 shows autocorrelator traces obtained before stretching and after compression. The small difference is due to different autocorrelators being used for the input and output pulses. Amplified pulses were also monitored on a single-shot autocorrelator. A typical autocorrelation for an amplified pulse of 500 μJ is shown in figure 1.8.

Measurements in the far field of the beam transmission through a pinhole indicated that the amplified beam of compressed pulses was about 1.8

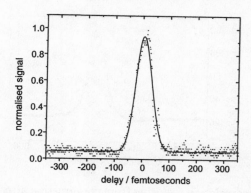

Fig. 1.8. A typical autocorrelation for an amplified pulse with an energy of 500 μJ per pulse. This autocorrelation was obtained on a single shot autocorrelator after amplification by the pre-amplifier only.

times the diffraction limit. Intensities in excess of 10^{16} W cm^{-2} would be expected with these parameters and this was confirmed by observing up to the seventh ionisation stage of argon in an ion time-of-flight spectrum [40].

The second amplifier is a conventional four-pass bow-tie configuration with a 10-mm-diameter, 7-mm-long Ti : sapphire crystal (Crystal Systems, FOM = 150 and $\alpha_{514} = 4.7$). An even number of passes is advisable since it limits the amount of astigmatism in the amplified beam. The crystal is pumped from both ends by two 532-nm beams from a Q-switched Nd : YAG laser (Spectra-Physics GCR) with a total energy of up to 850 mJ. The highest measured energy of amplified, uncompressed pulses from this amplifier was 150 mJ per pulse. However, the expected amplified energy is closer to 200 mJ per pulse, based upon the efficiency of Ti : sapphire. Our amplifier-modelling analysis indicates that the gain clamping near the 100-mJ level is due to parasitic modes in the rod, competing for gain at pump energies in excess of 700 mJ. In due course the rod will be clad with a material of suitable refractive index to minimise this problem. In the meantime the energy obtained is more than adequate for all current applications and allows the system to operate at the intended terawatt level.

Pulses of 50 fs are obtained after compression, as shown in figure 1.9. The FWHM of the spectrum of the amplified pulse, shown in figure 1.10, is sufficient to support a 46-fs Gaussian-shaped pulse. The fact that we are able to compress the stretched pulses from the Femto oscillator, i.e. without amplification, down to 32 fs suggests that the amplified pulses are subject to gain narrowing in the amplifier. This has also been observed by other developers of ultra-short-pulse amplifiers [41].

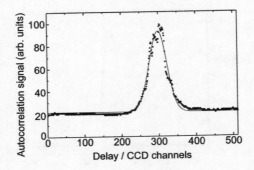

Fig. 1.9. An autocorrelation resulting from compression after the final amplifier. The full width at half maximum of the pulse is 50 fs.

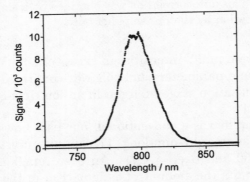

Fig. 1.10. The spectrum of the laser pulse shown in figure 1.9.

In order to obtain an estimate of the maximum intensity available from the terawatt laser, 5-mJ pulses of 50-fs duration were focused with an $f/2.5$ focusing mirror in an ionisation time-of-flight (TOF) spectrometer containing argon as a test gas. The TOF spectrum is shown in figure 1.11 and clearly shows the eighth stage of ionisation. This indicates that a focused intensity in the region of 10^{17} W cm^{-2} was obtained. This is consistent with the pulse parameters assuming a twice diffraction-limited beam. Since compressed pulse energies in excess of 50 mJ are available, the laser should be capable of providing a maximum focused intensity of at least 10^{18} W cm^{-2}.

1.4 Non-linear pulse-compression techniques

For a decade, the 'record' in pulse shortening was held by Fork *et al.* [42], who obtained 6-fs pulses of a few nanojoules by coupling 65-fs, 60-µJ pulses, at a centre wavelength of 620 nm, into a 9-mm-long fibre. The ultra-short pulses could be amplified to several microjoules, using dye

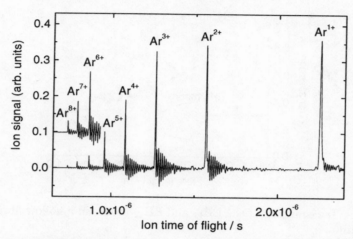

Fig. 1.11. The time-of-flight spectrum of argon. Courtesy of Dr Roy Newell's group from University College London.

mixtures, but at the expense of lengthening the pulse to 16 fs [43]. The use of single-mode fibres limits this technique to the damage threshold of the material.

Another technique for pulse shortening, which has recently appeared in the literature, is optical parametric amplification. Broad-bandwidth radiation from a white-light continuum source is amplified in a non-linear crystal, using the technique of parametric amplification [44]. Energies up to 1 μJ and a pulse duration of 11 fs between 500 and 700 nm at a repetition rate of 1 kHz have been produced [45]. Also this technique is limited by the damage threshold of the non-linear materials. Possibly, chirped pulse amplification can be used in combination with OPA technology to generate high-energy pulses at 10 Hz with a pulse duration as short as 20 fs [9].

The most advanced pulse-compression technique to date was developed in the second half of the 1990s. It is based on the spectral broadening inside a hollow noble-gas-filled fibre. It is now possible to use millijoule inputs and obtain up to 100 μJ outputs [46–48] at 800 nm, with pulses as short as 5 fs. A recent publication by Nibbering et al. [49] has extended the wavelength regime by compressing 50-fs, 400-nm pulses to less than 20 fs. Meanwhile Durfee et al. [50] have presented data on the use of hollow fibres as frequency-mixing media and showed that one can obtain 266-nm pulses from 800- and 400-nm inputs. Whilst their output spectrum has sufficient bandwidth to carry a 4.5-fs-duration pulse, it has not yet been compressed. Also presented in the paper is the extension of the technique to a tunable ultra-short femtosecond source by parametric amplification.

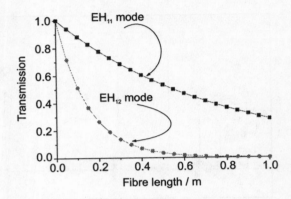

Fig. 1.12.   Transmission of the EH_{11} and EH_{12} modes in a hollow fibre of diameter 120 μm.

Propagation and modes in hollow fibres

Hollow fibres allow the exploitation of a spatially uniform, self-phase modulation (SPM) in an intense laser beam. Propagation along the fibre is mediated by grazing-incidence reflections at the dielectric–gas interface. Because of the many reflections, multi-modes cannot propagate through a long length of fibre. Marcatilin and Schmeltzer [51] showed that, if the refractive index between the gas and dielectric is less than 2.02, the two most likely modes to be transmitted are the hybrid EH_{11} and EH_{12} modes. The intensity profile as a function of the radial co-ordinate, r, is given by

$$I(r) = I_0 J_0^2 \left(\frac{2.405\, r}{a} \right) \qquad (1.15)$$

for the EH_{11} mode and

$$I(r) = I_0 J_0^2 \left(\frac{5.520\, r}{a} \right) \qquad (1.16)$$

for the EH_{12} mode, where I_0 is the peak intensity, J_0 is the zeroth-order Bessel function and a is the capillary radius. For these modes the phase constant, β, and the field-attenuation constant, α, are given by

$$\beta = \frac{2\pi}{\lambda} \left[1 - \frac{1}{2} \left(\frac{2.405\lambda}{2\pi a} \right)^2 \right], \qquad (1.17)$$

$$\frac{\alpha}{2} = \left(\frac{2.405}{2\pi} \right)^2 \frac{\lambda^2}{2a^3} \left(\frac{\nu^2 + 1}{\sqrt{\nu^2 - 1}} \right) \qquad (1.18)$$

for the EH_{11} modes and

$$\beta = \frac{2\pi}{\lambda} \left[1 - \frac{1}{2} \left(\frac{5.520\lambda}{2\pi a} \right)^2 \right], \qquad (1.19)$$

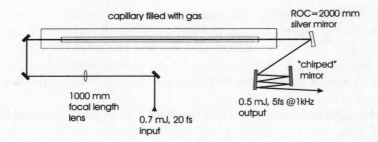

Fig. 1.13.   The experimental arrangement for a hollow-fibre experiment.

$$\frac{\alpha}{2} = \left(\frac{5.520}{2\pi}\right)^2 \frac{\lambda^2}{2a^3} \left(\frac{\nu^2 + 1}{\sqrt{\nu^2 - 1}}\right) \tag{1.20}$$

for the EH_{12} modes, where ν is the ratio of the refractive indices of the external (in this case fused silica) and internal (gas) media. Figure 1.12 shows the transmission of the EH_{11} and the EH_{12} modes through a 120-μm fibre. Clearly, as long as the fibre has sufficient length, discrimination against the higher EH_{12} mode can be achieved and the fibre will have single-mode transmission. Misoli *et al.* [46] used a 20-fs, 1-kHz system at 800 nm to introduce SPM in a 70-μm-diameter hollow fibre filled with Ar gas. They injected 80 μJ and achieved an output of 40 μJ with a pulse duration of about 5.2 fs. Higher energies can be achieved but it was noted that the wings of the autocorrelation trace increased. Pulse compression was achieved with the use of a pair of prisms and 'chirped' mirrors. The prisms were 2 m apart and had a small apex angle of 20°. The small prism angle reduces the second- and higher-order dispersion terms, which result from propagation of the laser pulse through the bulk media.

More recently, Sartania *et al.* [48] used only multiple reflections from 'chirped' mirrors to correct for group velocity dispersion and higher-order phase terms. In prisms, the high intensities induce self-focusing, which limits the maximum output. The hollow fibre was made of fused silica, was 260 μm in diameter, 850 mm long and filled with about 380 Torr of Ar. The experimental arrangement is shown in figure 1.13. With a laser input of 20 fs and 1 mJ, they achieved 500-μJ pulses of 5 fs, after compression. The measured beam profile was nearly diffraction limited ($M^2 = 1.08$ in the x-direction and $M^2 = 1.04$ in the y-direction). According to the authors, it should be possible to focus this beam to an intensity of greater than 10^{17} W cm^{-2}.

Extending the technique to 400-nm light, Nibbering *et al.* [49] used 100-μJ input pulses with a duration of 50 fs. They filled a 700-mm-long (ø = 110 μm) fibre with 2660 Torr of Ar and obtained an output of about

20 µJ before compression. The pulse was then compressed with 160-lines mm^{-1} gratings with a grating distance of 310 mm and an incidence angle of 2.4°. The output energy of 8 µJ and the pulse duration of 20 fs were measured with a FROG system. A higher pulse energy is possible if a pair of prisms is being used, but then the higher-order dispersion terms lead to a pulse duration of 24 fs.

Broad-bandwidth, ultra-short, ultraviolet pulses were obtained by Durfee *et al.* [50], who focused both 400 and 800 nm into a 127-µm diameter, 600-mm-long fibre filled with either argon or krypton. For inputs of 30 µJ at 400 nm and 64 µJ at 800 nm they achieved a 4 µJ output at 267 nm, corresponding to a 13% conversion efficiency with respect to the 400-nm pulse. The group did not try to compress the pulse but the bandwidth was sufficient to carry a 4.5-fs pulse. Using the technique of frequency mixing or optical parametric generation, they further showed that it is possible to tune the laser system through the ultraviolet-wavelength regime, with pulse durations as short as 10–20 fs.

Scaling to higher powers

The two most important considerations that have to be taken into account, when one is scaling to higher powers, are the critical power at which self-focusing occurs and the multi-photon ionisation threshold.

1. *Critical power.* The peak laser intensity must be smaller than the critical power for self-focusing. For a Gaussian beam profile, the critical power is given by $\lambda^2/(2\pi n_2)$, where n_2 is the non-linear refractive index of the gas. This parameter sets a constraint on the type of gas that can be used and its pressure.

2. *Multi-photon ionisation.* The peak power should be less than the multi-photon ionisation threshold of the gas. Christov *et al.* [52] recently showed that this threshold increases with decreasing pulse duration.

Thus, by a judicious choice of the type and pressure of gas and diameter of fibre and using the shortest possible input pulses, it would seem possible to scale the current technology to millijoule pulses at 10 fs.

1.5 Detection and determination of the pulse duration

Once ultra-short pulses have been generated and amplified to high energies, it is important to characterise them. Traditional autocorrelation by second-harmonic generation is usually unsatisfactory. Owing to the broad bandwidth of the ultra-short laser pulses, this method is hampered by the phase-matching conditions in the non-linear crystal. Besides, for many

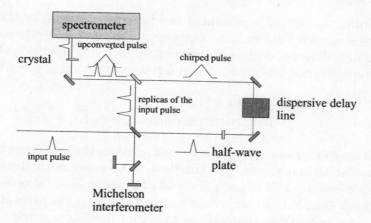

Fig. 1.14.   A schematic diagram of a real-time SPIDER apparatus.

high-field experiments accurate knowledge not only of the pulse shape but also of its phase is required. Let us briefly review a couple of techniques that can be used to recover the required information.

Frequency-resolved optical gating (FROG)

The FROG measurement is a spectrally resolved autocorrelation, whereby the autocorrelation is done in any material capable of providing a suitable non-linear process. Traditionally, FROG measurements have been obtained either by second-harmonic generation (SHG) or by using a polarisation gate (PG) [53]. The former is relatively easy to set up, but provides ambiguous time-direction information on the pulse. The latter technique, on the other hand, requires the introduction of polarisers, whose higher-order dispersion terms may affect the characteristics of the pulse being measured.

Another technique, transient-grating frequency-resolved optical gating (TG-FROG) [54] removes these problems. Here, two identical beams are overlapped in fused silica, to generate a transient grating. A third beam is diffracted by the grating to produce a fourth beam, which is then measured as the signal. This method overcomes the phase-matching problem and, since it uses a third-order process, it provides direction-in-time information that is not obtainable from second-harmonic autocorrelations. TG-FROG has been used successfully by Rundquist *et al.* [55] to characterise a 9-fs, 800-nm pulse.

Spectral phase interferometry

Spectral phase interferometry for direct electric-field reconstruction (SPIDER) is a recently introduced technique for the characterisation of ultrashort pulses [56]. SPIDER is able to retrieve the phase and amplitude of a pulse from the optical spectrum of two interfering pulses, by the use of a fast, non-iterative algorithm. The two pulses are identical except for their slightly different central frequencies. Under this condition they are said to be 'spectrally sheared'. The interferogram signal is given by

$$
\begin{aligned}
S(\omega) = {} & |E(\omega)|^2 + |E(\omega + \Omega)|^2 \\
& + 2 \, |E(\omega) E(\omega + \Omega)| \cos[\phi(\omega + \Omega) - \phi(\omega) + \omega\tau],
\end{aligned}
\tag{1.21}
$$

where Ω is the frequency or spectral shear, that is, the difference between the two central frequencies, τ is the time delay between the two pulses, $E(\omega)$ is the electric field of the pulse and $\phi(\omega)$ is the spectral phase of the pulse. Since the SPIDER algorithm recovers not only $E(\omega)$ but also $\phi(\omega)$ [56, 58], it is possible to characterise the input pulse completely.

A typical SPIDER apparatus is shown in figure 1.14. A portion of the input pulse is picked off by a beam splitter and directed through a Michelson interferometer, resulting in a pair of test pulses separated in time by a fixed delay τ (other experimental arrangements make use of an etalon to generate the pair of pulses [56]). The laser beam that is transmitted through the beam splitter, on the other hand, is being chirped by either a pair of gratings, in double pass (see equation (1.12)), or a block of material [57]. A waveplate rotates the polarisation of the chirped beam by 90°. The two perpendicularly polarised beams are recombined at another beam splitter and directed towards a type-II, frequency-doubling crystal. Since the two test pulses occur at different times, they interact with different frequencies of the chirped beam. The 'doubling' in the crystal thus generates two spectrally sheared pulses. Inside the spectrometer, the dispersed frequency components of the two pulses overlap in time and will therefore interfere. In other words, the spectrometer records an interferogram in the frequency domain. From the interferogram the spectral phase can be retrieved with certain mathematical algorithms [56, 58] that are based on fast Fourier transformations and filtering. Since SPIDER is a non-iterative phase-retrieval technique with a simple geometry and no moving parts, it is a very practical diagnostic and suitable for most high-repetition-rate laser systems.

1.6 Conclusions

It appears that femtosecond laser technology is reaching a mature stage. More and more scientists can now apply ultra-short-pulsed and/or superintense lasers in their experiments, thanks to the affordability and userfriendliness of all-solid-state laser systems. Naturally, such qualifications also appeal to those with a more pragmatic inclination and the use of this technology in applied areas is therefore likely to increase. In this first chapter, we have started by reviewing two very noteworthy developments, namely the chirped pulse-amplification technique and the self-mode-locked Ti:sapphire laser. As a practical example, we then presented the ASTRA system at the laser facility of the Rutherford Appleton Laboratory. Subsequently, we briefly discussed some non-linear pulse-compression techniques and found that particularly the use of gas-filled, hollow fibres appears very promising for more widespread applications. We finally concluded that 'FROG' and 'SPIDER' are ingenious and useful diagnostic systems for retrieving not only pulse length but also phase information.

Acknowledgements

The following are acknowledged for their support: the EPSRC for funding the programme, Dr Roy Newell's group at UCL for providing the argon-ionisation measurements, and all the users of the CLF's femtosecond facility who supported the upgrading of our facility.

References

[1] D. Strickland and G. Mourou, *Opt. Commun.* **56** 219 (1985).

[2] D. E. Spence, P. N. Kean and W. Sibbett, *Opt. Lett.* **16** 42 (1991).

[3] L. Xu, C. Spielmann, F. Krausz and R. Szipöcs, *Opt. Lett.* **21** 1259 (1996).

[4] S. Backus, J. Peatross, C. P. Huany, M. M. Murane and H. C. Kaptyn, *Opt. Lett.* **20** 2000 (1995).

[5] K. Yamakawa, M. Aoyama, S. Matsuoka, T. Kase, Y. Akahave and H. Takum. *Opt. Lett.* **23** 1468 (1998).

[6] M. Nisoli, S. Desilvestri and O. Svelto, *Appl. Phys. Lett.* **68** 2793 (1996).

[7] M. D. Perry *et al. Opt. Lett.* **24** 160 (1999).

[8] C. N. Danson *et al. J. Mod. Opt.* **45** 1653 (1998).

[9] I. N. Ross, P. Matousek, M. Towrie, A. J. Langley and J. L. Collier, *Opt. Commun.* **144** 125 (1997).

[10] A. Modena, Z. Najmudin, A. E. Dangor, C. E. Clayton, K. A. Marsh, C. Joshi, V. Malka, C. B. Darrow, C. N. Danson, D. Neely and F. N. Walsh, *Nature* **377** 606 (1995).

[11] M. Borghesi, A. J. MacKinnon, L. Barringer, R. Gaillard, L. A. Gizzi, C. Meyer and O. Willi, *Phys. Rev. Lett.* **78** 897 (1997).

[12] See for example: K. W. D. Ledingham *et al. Phys. Rev. Lett.* **84** 899 (2000); K. W. D. Ledingham and P. A. Norreys, *Contemp. Phys.* **40** 367 (1999); P. A. Norreys *et al. Phys. Plasmas* **6** 2150 (1999).

[13] J. H. Posthumus, J. Plumridge, L. J. Frasinski, K. Codling, A. J. Langley and P. F. Taday, *J. Phys. B* **31** L985 (1998).

[14] D. J. Smith, K. W. D. Ledingham, R. P. Singhal, H. S. Kilic, T. McCanny, A. J. Langley, P. F. Taday, C. Kosmidis, *Rapid Commun. Mass Spectrom.* **12** 813 (1998).

[15] See for example: J. Bonse, M. Geuss, S. Bandach, S. Sturm and W. Kautek, *Appl. Phys. A* **69** S399 (1999); X. B. Liu and X. L. Chen J. *Laser Applications*, **11** 268 (1999); J. Kruger and W. Kautek, *Laser Phys.* **9** 30 (1999).

[16] L. Xu, G. Tempea, Ch. Spielmann, F. Krausz, A. Stingl, K. Ferencz and S. Takano, *Opt. Lett.* **23** 789 (1998).

[17] X. Zhu, J.-F. Cormier and M. Piche, *J. Mod. Opt.* **43** 1701 (1996).

[18] J. C. Diels and W. Rudolph, *Ultrashort Laser Pulse Phenomena*, Academic Press, San Diego, 1996, ISBN 0-12-215492-4, p 317.

[19] S. Uemura and K. Torizuka *Opt. Lett.* **24** 780 (1999).

[20] C. Honninger, F. M. Genoud, M. Moser, U. Keller, L. R. Brovelli and C. Harder *Opt. Lett.* **23** 126 (1998).

[21] H. W. Kogelnik, E. P. Ippen, A. Dienes and C. V. Shank, *IEEE J. Quant. Electron.* **QE8** 373 (1972).

[22] B. Proctor and F. Wise, *Opt. Lett.* **17** 1295 (1992).

[23] A. Stingl, C. Spielmann, F. Krausz, and R. Szipöcs, *Opt. Lett.* **19** 204 (1994).

[24] O. E. Martinez, *IEEE J. Quant. Electron.* **QE23** 1385 (1987).

[25] M. Pessot, P. Maine and G. Mourou, *Opt. Commun.* **62** 419 (1987).

[26] E. B. Treacy, *IEEE J. Quant. Electron.* **QE5** 454 (1969).

[27] E. T. J. Nibbering, G. Gillon, M. A. Franco, B. S. Prade and A. Mysyrowicz, *J. Opt. Soc. Am. B* **14** 650 (1997).

[28] M. J. Shaw, C. J. Hooker and D. C. Wilson, *Opt. Commun.* **103** 153 (1993).

[29] E. T. J. Nibbering, M. A. Franco, B. S. Prade, G. Gillon, C. LeBlanc and A. Mysyrowicz, *Opt. Commun.* **119** 479 (1995).

[30] H. Li, F. Zhou, X. Zhang and W. Ji, *Opt. Commun.* **144** 75 (1997).

[31] B. E. Lemoff and C. P. J. Barty, *Opt. Lett.* **18** 1651 (1993).

[32] D. Du, J. Squier, S. Kane, G. Korn, G. Mourou, C. Bogusch and C. T. Cotton, *Opt. Lett.* **20** 2114 (1995).

[33] G. Cheriaux, P. Rousseau, F. Salin, J. P. Chambaret, B. Walker and L. F. Dimauro, *Opt. Lett.* **21** 414 (1996).

[34] A. Offner, *USA patent* 3,748,015 (1971).

[35] J. Itatani, Y. Nabekawa, K. Kondo and S. Watanabe, *Opt. Commun.* **134** 134 (1997).

[36] K. Yamakawa, P. H. Chiu, A. Magana and J. D. Kmetec, *IEEE J. Quant. Electron.* **QE30** 2698 (1994).

[37] M. Lenzner, Ch. Spielmann, E. Eintner, F. Krausz and A. J. Schmidt, *Opt. Lett.* **20** 1397 (1995); S. Backus, J. Peatross, C. P. Huang, M. M. Murnane and H. C. Kapteyn, *Opt. Lett.* **20** 2000 (1995); M. Li and G. N. Gibson, *J. Opt. Soc. Am. B* **15** 2404 (1998).

[38] Ch. Spielman, A. Stingl, R. Szipöcs and F. Krausz, *Opt. Lett.* **19** 204 (1994).

[39] I. T. McKinnie, A. C. Oien, D. M. Warrington, P. N. Tonga, L. A. W. Gloster and T. A. King, *IEEE J. Quant. Electron.* **QE33** 1221 (1997).

[40] S. Augst, D. Strickland, D. D. Meyerhofer, S. L. Chin and J. H. Eberly, *Phys. Rev. Lett.* **63** 2212 (1989).

[41] A. Antonetti, F. Blasco, J. P. Chamberet, G. Cheriaux, G. Darpentigni, C. LeBlanc, P. Rousseau, S. Ranc, G. Rey and F. Salin, *Appl. Phys. B* **65** 197 (1997).

[42] R. L. Fork, C. H. Cruz, P. C. Becker and C. V. Shank, *Opt. Lett.* **12** 483 (1987).

[43] G. Boyer, M. Franco, J. P. Chambaret, A. Migus, A. Antonetti, P. Georges, F. Salin and A. Brun, *Appl. Phys. Lett.* **53** 823 (1988).

[44] P. Matousek, A. W. Parker, P. F. Taday, W. T. Toner and M. Towrie, *Opt. Commun.* **127** 307 (1996).

[45] G. Cerullo, M. Nisoli and S. Desilvestri, *Appl. Phys. Lett.* **71** 3616 (1997).

[46] M. Nisoli, S. Desilvestri and O. Svelto, *Appl. Phys. Lett.* **68** 2793 (1996).

[47] M. Nisoli, S. Stagira, S. Desilvestri, O. Svelto, S. Sartania, Z. Cheng, M. Lenzner, C. Spielmann and F. Krausz, *Appl. Phys. B* **65** 189 (1997).

[48] S. Sartania, Z. Cheng, M. Lenzner, G. Tempea, Ch. Spielmann, F. Frausz and K. Ferencz, *Opt. Lett.* **22** 1 (1997).

[49] E. T. J. Nibbering, O. Duhr and G. Korn, *Opt. Lett.* **22** 1335 (1997).

[50] C. G. Durfee, S. Backus, M. M. Murnane and H. C. Kapteyn, *Opt. Lett.* **22** 1565 (1997).

[51] E. A. J. Marcatilin and R. A. Schmeltzer, *Bell Systems Techn. J.* **43** 1783 (1964).

[52] I. P. Christov, J. Zhou, J. Peatross, A. Rundquist, M. M. Murnane and H. C. Kapteyn, *Phys. Rev. Lett.* **77** 1743 (1996).

[53] K. W. DeLong, R. Trebino and D. J. Kane, *J. Opt. Soc. Am. B* **11** 1595 (1994).

[54] J. M. Sweetser, D. N. Fittinghoff and R. Trebino, *Opt. Lett.* **22** 510 (1997).

[55] A. Rundquist, C. Durfee, Z. Chang, G. Taft, E. Zeek, S. Backus, M. M. Murnane, H. C. Kaptyn, I. Christov and V. Stoev, *Appl. Phys. B* **65** 161 (1997).

[56] T. M. Shuman, M. E. Anderson, J. Bromage, C. Iaconis, L. Waxer and I. A. Walmsley, *Opt. Express* **5** 134 (1999).

[57] L. Gallmann, D. H. Sutter, N. Matuschek, G. Steinmeyer, U. Keller, C. Iaconis and I. A. Walmsley, *Opt. Lett.* **24** 1314 (1999).

[58] C. Iaconis and I. A. Walmsley, *IEEE J. Quant. Electron.* **QE35** 501 (1999).

2

Diatomic molecules in intense laser fields

Jan H. Posthumus[†]

The Blackett Laboratory, Imperial College
Prince Consort Road, London SW7 2BQ, England, UK

James F. McCann

School of Physics and Mathematics
Queen's University
Belfast BT7 1NN, Northern Ireland, UK

[†] *Present address: Sektion Physik, Universität München*
Am Coulombwall 1, D-85748 Garching, Germany

2.1 Introduction

When an intense laser pulse passes through a gas, the laser–matter interactions are highly non-linear and lead to extensive changes both in the nature of the transmitted light and the medium. Even if the excitation frequencies of the molecule are not in resonance with that of the light, the external electric field can exceed the internal binding forces and allow strong absorption of energy. Subsequently the energy is dissipated through explosive fragmentation of the molecule and by emission of high-frequency light. The physics of strongly correlated many-body quantum systems interacting with intense dynamic external fields is extremely complicated. The understanding of such processes in simple atoms is still in development, so it is fair to say that the mechanisms of multi-electron photodissociative ionisation of molecules are still far from being understood. Naturally the physical process of electron removal is very similar for both types of system. Indeed, the growth in interest in molecular dynamics in intense fields was originally fuelled by speculation on the character of multi-electron ejection from atoms [1]. One of the topical issues of debate in molecular physics focuses on the sequence of the fragmentation; whether the electrons are liberated sequentially or simultaneously and how these processes depend on the nuclear motion. In light atoms the electron correlation in the outer shells is important irrespective of whether the electrons emerge sequentially, whereby the electrons are peeled off the atoms one by one, or escape in groups arising from a multiple collective excitation of the system. Whichever is the case, experiments on molecules led to the possibility of performing detailed particle spectroscopy on the ejected electrons and ions, resolving energy and angular dependences. The

27

explosive process of photodissociative ionisation, which is characteristic of such interactions, produces a hot plasma of fast electrons and ions, which can interact to generate high-energy photons [2, 3]. If this heating process occurs within an atomic cluster [4, 5], the energy can be harnessed to such an extent that nuclear fusion can occur. Determining the physics of the molecule–laser interaction is a key step towards understanding the complex phenomena arising in atomic clusters.

Analysing atomic dynamics within molecular chains requires knowledge of motions on the timescale of molecular vibrations, termed *ultra-fast processes* in condensed matter physics and biology. Consequently, the discovery and development of light sources that can probe such processes is of great interest. The shortest pulses produced nowadays are already somewhat shorter than the vibrational period of the hydrogen molecule ($\simeq 10^{-14}$ s). Conventional low-intensity CW lasers are primarily concerned with high-resolution energy spectroscopy. However, with pulsed lasers one can study and influence the dynamics of vibrational modes of atoms within molecules and clusters in the time domain.

Multi-photon ionisation at very high laser intensities can deplete the initial state in a few tens of femtoseconds and can therefore not be studied properly over picosecond timescales, corresponding to bending modes of vibration or rotational excitation. Such long and intense pulses would saturate the ionisation process, thus making it impossible to measure the rate of ionisation. The intensity must rise quickly so that the initial state survives the leading edge of the pulse envelope. The application of new techniques in pulse compression will undoubtedly lead to the discovery of novel high-intensity phenomena.

Photodissociation of a molecule without ionisation is the predominant physical process at low light intensities. Light diatomic molecules such as H_2 have very low polarisabilities and are therefore difficult to excite by non-resonant radiation. In the vacuum-ultraviolet wavelength range, the highly excited molecular states are accessible and 'resonance-enhanced multi-photon ionisation' (REMPI) of H_2 has been studied with picosecond and nanosecond laser pulses. Pulse-length effects will become apparent in the femtosecond regime, which should soon be available with tunable, short-wavelength pulses. This will lead to new opportunities to study and influence the dynamics of multi-photon excitation of H_2 and its isotopes HD and D_2.

This chapter concentrates particularly on femtosecond laser pulses of the Ti : sapphire laser ($\lambda \simeq 740$–950 nm). Experiments reveal that homo-nuclear diatoms subjected to intensities in excess of 10^{14} W cm^{-2} invariably ionise before they dissociate. As soon as the molecule has been ionised, its polarisability increases enormously due to *charge resonance* [6]. The inversion symmetry and open-shell configuration mean that the molecular orbitals of *gerade* and *ungerade* symmetry are readily mixed by the external electric field. In H_2^+, for example, only one of the two 1s atomic orbitals is occupied and the shift of the single electron towards one atomic core therefore requires little excitation energy. When this happens, the molecular bond is broken and the ion dissociates: $H_2^+ \rightarrow H + H^+$. Section 2.4 is devoted to the sequential dissociative ionisation of H_2 and discusses especially the dynamics of the strong-field dissociation of the one-electron molecule H_2^+.

Multiple ionisation of molecules is inherently difficult to analyse by particle spectroscopy owing to the variety and large number of fragments. In large molecules and clusters, which contain many, virtually equivalent electrons, laser-induced collective excitations and multi-electron interactions will prevail and classical theories or many-body mean-field quantum theories, such as density-functional theory [7], must take this into account. However, in the case of diatomic molecules, the multiple ionisation can be quite successfully treated as a sequence of independent over-the-barrier electron-releasing processes. Indeed, the field-ionisation, Coulomb-explosion (FICE) model, which is discussed in section 2.5, can account for a great deal of the data observed to date.

One of the most interesting aspects of the strong-field ionisation of molecules is the enhancement of the ionisation rate at critical distances. The enhancement is discussed in terms of three approaches, namely the classical FICE model and both a quasi-static and a time-dependent quantum-mechanical calculation. The relationship among the results of the three methods is also critically analysed in section 2.5.

The subject of molecular and cluster dynamics in intense laser fields is still evolving rapidly and many aspects are as yet poorly understood. The scope of this chapter does not permit us to discuss, for example, the highly interesting subject of laser-induced alignment of the molecular axis and rotational heating effects. However, a concise selection of follow-up reading can be found in the concluding section (2.6) of this chapter.

2.2 Experimental considerations

In the photodissociation of diatomic molecules by lasers of moderate intensities ($< 10^{14}$ W cm^{-2}) [8–10] ionisation rates are usually very small. This is simply explained by the fact that the external electric field is much weaker than the internal fields which bind the electrons. When the wavelength is as long as $\lambda = 10.6$ μm and the laser intensity is higher, the laser acts as a quasi-static electric field that removes electrons through the process of tunnelling. This process can occur before dissociation of the molecular ion [11]. (However, it is not a general feature insofar as the opposite is often true for the more loosely bound triatoms [12].) It is possible, therefore, to study the kinetic-energy spectra of the electrons with, for example, a magnetic-bottle spectrometer [13], but in general the ionic fragments carry more information on the dynamics of the dissociation process. Since pulsed lasers are being used, the kinetic energy of the ions can be measured by the time-of-flight technique [14].

Figure 2.1 is a schematic diagram of a typical experimental set-up. The laser beam is focused inside a high-vacuum chamber, where it interacts with a target gas. The lower extraction electrode has a relatively small opening through which the ions enter a field-free drift zone before they hit the multi-channel plates of the detector. As a result of the small aperture in front of the detector, only ions that are ejected approximately parallel to the axis of the drift tube reach the detector. Each fragmentation channel thus produces two peaks in the time-of-flight spectrum, corresponding to 'forward' ions ejected towards the detector and 'backward' ions ejected away from the detector but whose trajectories are reversed by the extraction field. The small angular acceptance of the aperture of the drift tube ensures that an accurate correlation between the time of flight of an ion and its initial kinetic energy is obtained. A typical time-of-flight spectrum of the strong-field ionisation of H_2 is shown in figure 2.2. The height of the 'backward' peak is lower than that of the 'forward' peak as a result of the wider angular dispersion of the ions before collection. The angular distributions of the fragment ions with respect to the laser polarisation are found by rotating the plane of polarisation of the light. A larger opening and a shorter drift tube can be used to enlarge the acceptance angle and hence improve the count rate. Correlation measurements, as described in chapter 3, benefit from a high collection efficiency [16–18]. A strong extraction field that nullifies the initial kinetic energies further improves measurements of the total ion yield as a function of laser intensity [19, 20].

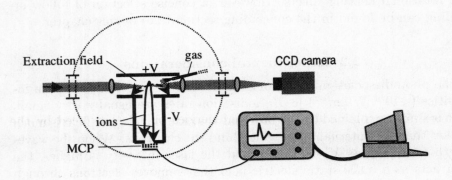

Fig. 2.1. A schematic diagram of a typical experimental set-up for the study of intense-laser–molecule interactions by ion time-of-flight spectroscopy.

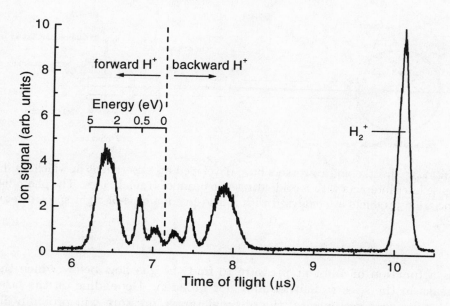

Fig. 2.2. A typical ion time-of-flight spectrum of the strong-field ionisation of H_2, showing the H_2^+ peak and forward and backward protons (adapted from Thompson *et al.* [15]).

Target gas can be introduced anywhere in the vacuum system, but it will be used most efficiently when it enters near the focus. At a given target density, this will also reduce the ambient density, which could lead to unwanted collisions. However, the proximity of gas inlet and interaction region is severely limited by the requirement for a uniform ion-extraction field.

At room temperature, Doppler broadening restricts the energy resolution, particularly for ions whose energies lie below 0.5 eV [21]. However, thermal broadening can be minimised by the use of a molecular beam. Rottke *et al.* [22] recently developed a very elegant apparatus with a supersonic, cold H_2 beam without an extraction field. A schematic diagram of their experimental set-up is shown in figure 2.3. The absence of an extraction field ensures that the angular acceptance of the spectrometer is not dependent on the kinetic energies of ions.

When one is measuring the kinetic energies of ions produced by the interaction of laser radiation with isolated molecules, it is important to avoid space-charge effects. When more than one molecule in the interaction region is ionised, the emerging ions can exchange energy with each other, leading to a distortion of the time-of-flight spectrum. As long as the average separation between the target molecules is fairly large, these collision effects can be neglected. In practice, the ion spectra are studied

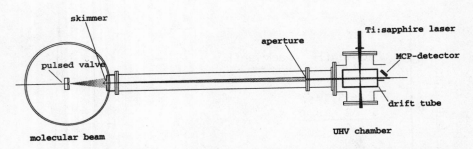

Fig. 2.3. Rottke and coworkers have developed an experiment in which an intense laser interacts with a cold, ultrasonic beam of H_2 molecules. The photodissociation products are analysed with a field-free ion time-of-flight spectrometer [22].

as a function of ambient pressure to find the gas flow below which the peaks in the spectra neither shift nor broaden. Depending on the laser intensity and gas species, the optimal target pressure will probably lie somewhere in the range 10^{-5}–10^{-7} mbar.

Even for these low pressures, ions originating from different molecules will often arrive simultaneously at the detector. The complete signal from the multi-channel plates (MCP) is, therefore, usually recorded with a digital oscilloscope. It is useful to store as many single-shot spectra as possible in the memory of the oscilloscope before transferring them to the computer and thus reduce the considerable 'deadtime' originating from the initialisation of the data transfer. The successive pulses produced by high-power lasers tend to have slightly different characteristics. A variation of 5% or 10% in laser intensity, which is typical, will strongly affect a highly non-linear process. Sorting the spectra according to the laser intensity, a process often called 'binning', can surmount some of these problems. Assuming that the pulse shape is reasonably stable, it suffices to record the pulse energies, for example with a photodiode that is connected to the second channel of the oscilloscope.

The method of acquiring all single-shot spectra can accommodate a laser repetition rate of 10 Hz, but higher repetition rates, e.g. 1 kHz, can nevertheless be beneficial for certain types of experiments. Correlation measurements are often time-consuming and benefit from a higher repetition rate. Refinements in the electronics such as the use of a multi-stop time-digital converter are required to record the coincidence events. More details on this can be found in chapter 3.

By examining the photodiode signal with a single-channel analyser, it is possible to trigger the oscilloscope only if the laser pulse falls within a narrow intensity window. At a repetition rate of 1 kHz it is practical even when a significant percentage of the laser pulses is discarded. The digital oscilloscope can take the average of a large number of spectra at the right intensity, which is then transferred to a PC. The number of spectra rejected will remain limited, if the laser is relatively stable.

Once the maximum allowable target density has been found, the signal of a certain process might still be very small, either because it is weak or because it is concealed by more dominant processes. Non-linear processes often occur within small intensity windows. Whilst a given minimum intensity is required to make the process efficient, the competing higher-order processes suppress it at higher intensities. When this is the case, the signal strength is optimised by maximising the focal volume of the specific intensity window. To this end, all the available laser power is used in combination with the longest possible focal length that still achieves the required laser intensity. Despite the larger volume, the target density needs to be reduced only slightly, because space-charge effects are particularly sensitive to the charge density, thus allowing an increase in the ion signal.

The volume of the focus

The high power densities at which strong field effects become prominent require strong focusing of the laser beam. The peak intensity can be estimated from the ratio of the peak power of the laser to the cross section of the focus. For example, a pulse of duration 100 fs with 100 µJ energy yields a peak power of approximately 10^9 W. Taking the beam to be in the Gaussian TEM_{00} mode, the beam radius, w, is defined by the expression [23]

$$I(r) = I_{\max} e^{-2r^2/w^2}, \tag{2.1}$$

where r is the radial distance from the beam axis. Focusing with a lens or parabolic mirror of focal length f yields a waist radius, w_0, of

$$w_0 = \lambda f / (\pi w). \tag{2.2}$$

So, for example, when a Ti:sapphire laser beam of 5 mm diameter (w = 2.5 mm, λ = 800 nm), is focused with a parabolic mirror with f = 2 cm, this yields $w_0 \approx 2$ µm. Thus the peak intensity at the focus is of the order of 10^{16} W cm^{-2}, which is comparable to the electric field in the hydrogen atom at the Bohr radius. In practice, imperfections in the quality of the beam and the optics prevent the focal spot from reaching the diffraction limit; the shorter the focal length the larger the discrepancy. When either a lens or an off-axis parabolic mirror is used, the focal properties and optics of the beam can be studied with a CCD camera using a microscope objective lens in front of the camera.

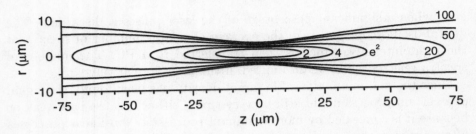

Fig. 2.4. Iso-intensity contours at the focus of a Gaussian laser beam. The labels refer to the ratio of the intensities at the centre of the focus and at the contour.

An ever-present technical difficulty is the spatial variation of the laser intensity around the focus, with different processes occurring simultaneously in various parts of the focal volume. The signal from a certain process will be proportional to the intensity dependent probability, $P(I)$, and the volume of the focal shell where this intensity pertains. For the Gaussian beam [23]

$$I = \frac{I_f}{1 + (z/z_R)^2} \exp\left(-\frac{2r^2}{w_0^2\left[1 + (z/z_R)^2\right]}\right), \tag{2.3}$$

where I_f is the intensity at the focal point and z is the axial coordinate. The Rayleigh range, z_R, is defined according to

$$z_R = \pi w_0^2/\lambda \tag{2.4}$$

and is the distance along the beam axis ($r = 0$) at which the intensity reduces to 50% of its maximum value. Using the example discussed earlier, $\lambda = 800$ nm and $w_0 \approx 2\mu$m, this gives us $z_R = 16$ μm. Inverting equation (2.3) for r at $I = I_i$ yields

$$r_i = \frac{w_0^2}{2}\left[1 + (z/z_R)^2\right]\left[\ln\left(\frac{I_f/I_i}{1 + (z/z_R)^2}\right)\right]^{\frac{1}{2}}, \tag{2.5}$$

which describes a contour surface (shell) of constant intensity I_i. The iso-intensity shells that are drawn in figure 2.4 are labelled by the ratios I_f/I_i. Thus the axial extent of shell '2' is equal to the Rayleigh range and the radial extent of shell 'e^2' at $z = 0$ is equal to w_0. Note that the vertical scale of figure 2.4 is twice as large as the horizontal.

The focal volume where the intensity is larger than I_i, is found from integrating the volume within the contour surface:

$$V(I > I_i) = \int\int 2\pi r\,dr\,dz = \int |\pi r^2|\,dz$$

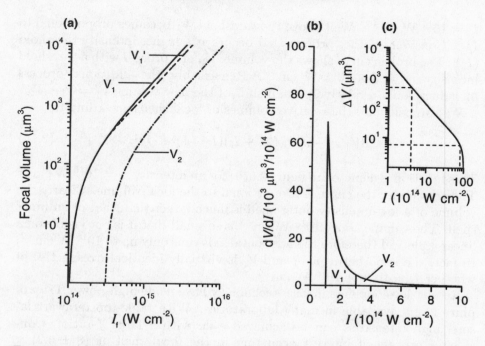

Fig. 2.5. (a) The focal volume for $I > 10^{14}$ W cm^{-2} as a function of I, see equation (2.8). (b) dV/dI for a focused intensity $I_f = 10^{16}$ W cm^{-2} as a function of I, see equation (2.10). (c) The volume of a thin shell with $dI/I = 0.05$, also for $I_f = 10^{16}$ W cm^{-2}.

$$= \int \pi w_0^2 \Big[1 + (z/z_R)^2\Big] \ln\!\Big(\frac{I_f/I_i}{1 + (z/z_R)^2}\Big) dz, \qquad (2.6)$$

The integration limit, z_i, is found from equation (2.3) at $r = 0$:

$$z_i = z_R \sqrt{\frac{I_f}{I_i} - 1}. \qquad (2.7)$$

Evaluating the integral yields the expression

$$V(I > I_i) = \pi w_0^2 z_R \Big(\frac{4}{3}\beta + \frac{2}{9}\beta^3 - \frac{4}{3}\arctan\beta\Big), \qquad (2.8)$$

with

$$\beta = \sqrt{(I_f - I_i)/I_i}. \qquad (2.9)$$

In figure 2.5(a) an example of the rapid expansion of the focal volume with respect to the intensity is illustrated. The solid curve shows V_1 when $P(I < I_1) = 0$ and $P(I > I_1) = 1$ with the threshold intensity

$I_1 = 10^{14}$ W cm^{-2}. Well above the threshold, V_1 becomes proportional to $I_{\mathrm{f}}^{1.5}$. Consider another process that occurs at a higher intensity threshold (I_2). The broken curve shows the volume, V_2, of a process with a threshold intensity $I_2 = 3 \times 10^{14}$ W cm^{-2}. Remarkably, the additional process introduces only a relatively small shift of the new volume $V_1' = V_1 - V_2$.

We can calculate the relative volumes of focal shells according to

$$|\mathrm{d}V| = \frac{\pi}{3} w_0^2 z_{\mathrm{R}} (2I + I_{\mathrm{f}})(I_{\mathrm{f}} - I)^{\frac{1}{2}} I^{-\frac{5}{2}} \, \mathrm{d}I. \qquad (2.10)$$

This function is depicted in figure 2.5(b) for an intensity $I_{\mathrm{f}} = 10^{16}$ W cm^{-2}. The area under the curve is proportional to the focal volume. Clearly the volume of a low-intensity outer shell is much larger than that of an inner shell. The volume above 10^{15} W cm^{-2} is so small that it is not visible on a linear scale and therefore the horizontal axis runs only up to 10^{15} W cm^{-2}. In fact, the ratio between V_1 and V_2 is virtually identical irrespective of whether $I_{\mathrm{f}} = 10^{15}$ or 10^{16} W cm^{-2}.

Recent developments in laser technology have made high-power Ti : sapphire lasers available in many laboratories. With table-top terawatt lasers, high intensities can be achieved even with a large f-ratio. Consider a process of interest occurring in the focal shell at $(8 \pm 0.2) \times 10^{15}$ W cm^{-2} that produces a signal in the time-of-flight spectrum very close to that of another process occurring at $(4 \pm 0.1) \times 10^{14}$ W cm^{-2}. Both these shells have $\mathrm{d}I/I = 0.05$. The volume of such shells is, to a good approximation,

$$\Delta V(\mathrm{d}I/I = 0.05) = \frac{\pi}{60} w_0^2 z_{\mathrm{R}} (2I + I_{\mathrm{f}})(I_{\mathrm{f}} - I)^{\frac{1}{2}} I^{-\frac{3}{2}}, \qquad (2.11)$$

and is plotted in figure 2.5(c), again for $I_{\mathrm{f}} = 1 \times 10^{16}$ W cm^{-2}. The curve shows that the volume of the shell at the lower intensity is 80 times larger than the volume at the higher intensity. This result is independent of the f-ratio. The signal of interest is, therefore, easily overwhelmed if the spectrometer samples the products from the whole range of intensities.

The modest decrease of the intensity along the axis of the laser beam is the determining factor for focal volumes at lower intensities. However, it is possible to screen off most ions from these outer shells and make the signals of processes at high and low intensity of comparable strength, particularly if one uses a large f-ratio.

Figure 2.6 illustrates a focus of the same beam as before, in which a lens of $f = 30$ cm is used. Note that the vertical scale of figure 2.6 is ten times larger than the horizontal scale. While the waist radius of the focus is only 30 μm, the Rayleigh range is now nearly 4 mm. With a small slit in the extraction electrode, only ions for which $z \ll z_{\mathrm{R}}$ can reach the detector. The intensity distribution of the unscreened part of the focus is

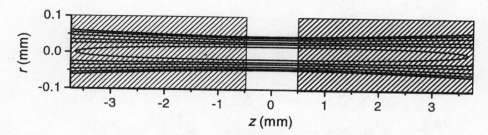

Fig. 2.6. Contours for $f = 30$ cm. Note the tenfold difference in the vertical and horizontal scales. Owing to the long Rayleigh range of almost 4 mm, large uninteresting parts of the focus can be screened off with a slit.

approximately

$$I \approx I_{\mathrm{f}}\, e^{-2r^2/w_0^2}. \tag{2.12}$$

The contours are now

$$r_{\mathrm{i}} = \sqrt{\frac{w_0^2}{2} \ln\!\left(\frac{I_{\mathrm{f}}}{I_{\mathrm{i}}}\right)} \tag{2.13}$$

and the focal volume has the form

$$V(I > I_{\mathrm{i}}) = l\pi r_{\mathrm{i}}^2 = \frac{\pi}{2} l w_0^2 \ln\!\left(\frac{I_{\mathrm{f}}}{I_{\mathrm{i}}}\right), \tag{2.14}$$

where l is the width of the slit. Figure 2.7(a) shows the sizes of the screened volumes V_1, V_2 and V_1' for $l = 1$ mm. As soon as a sequential process starts, the volume ceases to grow. Although the cylindrical shell steadily expands with increasing intensity, the larger radius is exactly compensated by a decrease in thickness of the shell. The differential volume $\mathrm{d}V$ is given by:

$$|\mathrm{d}V| = \frac{\pi}{2} l w_0^2 \frac{\mathrm{d}I}{I}, \tag{2.15}$$

which is plotted in figure 2.7 (b). Indeed, for $I_{\mathrm{f}} > I$, the volume of a thin shell is independent of the laser intensity, I_{f}. Furthermore, shells with equal relative thickness, $\mathrm{d}I/I$, are now equally large (see figure 2.7 (c)), which will make it much easier to observe signals from processes that occur in the centre of the focus.

The disadvantage of a long focal length is the need for a more powerful laser. In the above example f was increased by a factor of 15 and, in order to reach the same I_{f}, the pulse energy must now be 15^2 times larger, that is of the order 20 mJ.

The most effective screening design, which maintains a satisfactory collection efficiency, is obtained by placing the slit as close as possible to the focus. A reasonable choice would be a slit of width 1 mm at a distance of

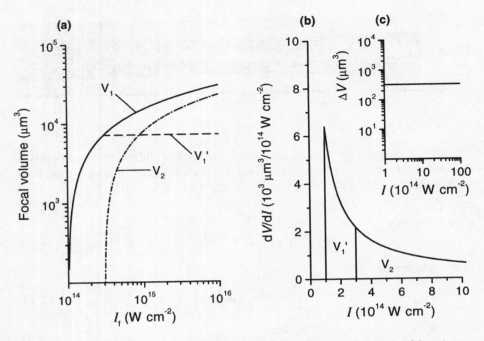

Fig. 2.7.   (a) The screened volumes V_1, V_2 and V_1' for a slitwidth of $l = 1$ mm, see equation (2.14). (b) dV/dI for $I_f = 10^{16}$ W cm^{-2}, see equation (2.15). The difference between the central and outer shell is clearly much less than that in the case of an unscreened focus, cf. figure 2.5(b). (c) With a slit in place, the visible parts of focal shells with the same dI/I are also equal in size, cf. figure 2.5(c).

1 mm from the focus. In order to limit distortions of the extraction field inside the slit, the thickness of the electrode should be considerably less than the diameter of the opening, e.g. 0.1 mm. After the first electrode, further acceleration of the ions is necessary. The potential benefits of this screening method were demonstrated by Hansch *et al.* [24]. It holds the promise of allowing further investigation of the correlation between intensity shells and Coulomb-explosion channels [25].

2.3 Appearance intensities

The most straightforward experiment in the present context is the study of the ion yield as a function of laser intensity. When H_2 is subjected to ultra-short laser pulses at $\lambda = 375$ nm, three distinct ion channels emerge at progressively higher intensities [15]. First the H_2^+ ions appear, then protons of energy $\simeq 0.5$ eV, due to the $H + H^+ + e^-$ channel labelled (0,1) and, finally, the more energetic protons from the $H^+ + H^+ + 2e^-$

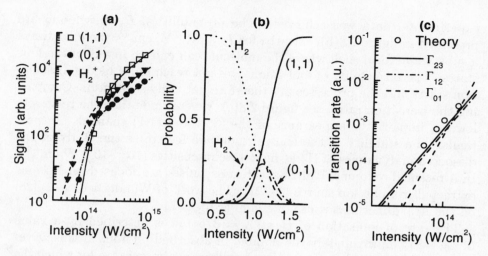

Fig. 2.8. (a) Log–log plots of ion signal versus intensity for H_2 subjected to 85-fs laser pulses at $\lambda = 375$ nm. The smooth curves result from a straightforward simulation of the sequential processes. (b) Population probabilities, P_i, and (c) the transition rates, Γ_{ij}, following the simulation described in the text. Adapted from [15].

channel, labelled (1,1). Figure 2.8(a) shows log–log plots of the signals of these channels versus the laser intensity. The results are simulated (smooth curves) assuming that the three processes are sequential. The rate equations for the populations, P_i, are

$$dP_0 = -\Gamma_{01}\, dt,$$
$$dP_1 = (\Gamma_{01} - \Gamma_{12})\, dt,$$
$$dP_2 = (\Gamma_{12} - \Gamma_{23})\, dt,$$
$$dP_3 = \Gamma_{23}\, dt, \qquad (2.16)$$

where Γ_{01} is the rate of ionisation of H_2, Γ_{12} is the rate of dissociation of H_2^+ and Γ_{23} is the final ionisation rate. Simple power laws, $\Gamma_{ij} = \alpha_{ij} I^{n_{ij}}$, are assumed to describe the transition rates. The equations (2.16) were solved analytically by approximating the laser pulses by square pulses of duration 50 fs. The initial conditions are that $P_0 = 1$ and $P_1 = P_2 = P_3 = 0$. In order to obtain the ion signals, S_i, at I_f, the intensity at the centre of the focus, the probabilities are weighed by the volumes of iso-intensity shells and then integrated numerically, i.e.

$$S_i(I_f) = \rho \int P_i \frac{dV}{dI}\, dI, \qquad (2.17)$$

where ρ is the target density and dV/dI is given by equation (2.10). The values for a_{ij} and n_{ij} were adjusted until a reasonably good fit with the experimental data was achieved. The probabilities, P, thus found are presented in figure 2.8(b). Clearly, for $I_0 \geq 10^{14}$ W cm^{-2}, there is always a substantial focal shell where the molecule can end up in any of the four states. The probability of ionisation does not reach unity before dissociation starts and the processes are therefore not truly sequential. Comparing the transition rates, see figure 2.8(c), we conclude that the relatively low Γ_{01} impairs the appearance of the (0,1) and (1,1) channels. In particular, the stability of the neutral molecule frustrates the study of the dissociation dynamics of H_2^+ at moderate intensities (10^{12}–10^{13} W cm^{-2}). In a recent development ultra-short laser pulses are focused on H_2^+ ions extracted from an ion source [26, 27]. The work of Wunderlich *et al.* [28] on the Ar_2^+ dimer makes use of a similar idea.

The rate of ionisation of H_2 at $\lambda = 375$ nm is described by a value of $n_{01} = 6$, approximately the number of absorbed photons at this wavelength. This is not the case for the two subsequent processes, for which the fits yield $n = 4$. They are thus definitely non-perturbative in character. At the longer wavelength of $\lambda = 750$ nm, $n_{01} = 6$ is found, which is again much lower than the minimum number of absorbed photons predicted by leading-order perturbation theory.

Whenever the ionisation potential is much larger than the photon energy, the character of intense laser ionisation can be loosely divided into different groups. The classification is guided by the value of the Keldysh parameter, which is a measure of the ratio of the electron energy in the atom to the energy in the field. It can be shown that an equivalent expression of this parameter is the ratio of the tunnelling time for field ionisation and the optical period. It is defined as [29]

$$\gamma = \sqrt{|E_i|/(2U_p)}, \qquad (2.18)$$

where $|E_i|$ is the ionisation potential and U_p the kinetic (ponderomotive) energy of a free electron in the laser field with zero drift velocity. The ponderomotive potential is given by

$$U_p = [F_0/(2\omega)]^2, \qquad (2.19)$$

where F_0 and ω are the amplitude and angular frequency of the laser electric field, all quantities being in atomic units. The conversion formula relating field strength and laser intensity is

$$F_0 \approx \left(\frac{\text{Intensity in W cm}^{-2}}{3.51 \times 10^{16}} \right)^{\frac{1}{2}}. \qquad (2.20)$$

Thus, at an intensity of 10^{14} W cm^{-2}, the ionisation of H_2 at $\lambda = 375$ nm is characterised by $\gamma \approx 2.5$. The condition $\gamma \gg 1$ corresponds to *multiphoton ionisation*, in which the electron acquires energy over many laser cycles, by gradually building up an oscillation in phase with the laser field. The action of scattering with the molecule is essential to create the anharmonic excitations which allow the electron to absorb the energy and momentum which eventually allow it to escape. The effectiveness of this mechanism is greatly enhanced if one of the harmonic frequencies of the electron oscillation coincides with a resonance.

The second mechanism is *tunnelling ionisation* [30–32], in which the probabilities of ionisation during each laser cycle are more or less independent. The electron does not interact with the nuclear centres but instead is drawn towards the edge of the molecular orbit, where tunnelling can occur. The electron emerges with a drift velocity that is determined by the momentum distribution of the initial state; no exchange of momentum is effected since scattering does not occur. At $\lambda = 750$ nm, $\gamma \approx 1$ and the laser acts as a quasi-static electric field that allows the electrons to escape from the parent molecule by tunnelling through a potential barrier. Finally, if the field is stronger than the internal binding field the electron can be drawn out without the need to overcome an energy barrier. This *over-the-barrier ionisation* regime for which $\gamma \ll 1$ is pertinent whenever the probability of ionisation approaches unity within one laser cycle [33].

2.4 The photodissociation of H_2^+ in intense laser fields

The semiclassical approximation

By definition an optical pulse cannot be shorter than one cycle. This corresponds to a pulse length of $\simeq 3$ fs for the Ti:sapphire laser. For a diffraction-limited spot, in order to achieve an intensity of the order of 10^{14} W cm^{-2}, the laser pulse must contain at least 1 nJ. Consequently, with 10^{10} equivalent photons in the immediate vicinity of the molecule, the laser field can be treated classically. The interaction with the magnetic component of the field is negligible as long as the electron velocities remain non-relativistic.

Although the laser field behaves classically, it nevertheless leads to quantized *multi-photon* processes. For example, in the process of above-threshold ionisation (ATI) [34] the electron-energy spectrum displays a series of equidistant peaks that are separated by the photon energy. However, this corresponds to frequency multiplication that occurs in an anharmonic oscillator, be it quantum or classical. The signature of the frequency multiplication is most clearly seen in the spectrum of emitted light of high-order harmonics of the incident light.

Time-dependent propagation of wavepackets

The photodissociation of H_2^+ in short-pulsed laser fields cannot be analysed properly with time-independent theories, since the laser intensity varies on the timescale of the dissociation process. Consequently, the time-dependent Schrödinger equation,

$$i\hbar\,\partial_t\Psi = (T_\mathrm{N} + H_\mathrm{e})\,\Psi, \tag{2.21}$$

needs to be integrated explicitly. In this notation the nuclear kinetic energy is denoted by T_N while the electronic Hamiltonian (including the molecule–laser interaction) is written as H_e.

Let us assume that the pulse is much shorter than the typical rotational period of the molecule, so that we need consider only nuclear vibrational relaxation. The associated electric field, within the dipole approximation, is denoted as $\boldsymbol{F}(t)$. The Born–Oppenheimer approximation proposes that the electronic and nuclear motions are weakly coupled and can be separated. Thus

$$\Psi(\boldsymbol{r}, R, t) = \sum_j \phi_j(\boldsymbol{r}, R)\chi_j(R, t) \tag{2.22}$$

and, on averaging over the electronic coordinate and neglecting the non-adiabatic dynamic couplings, we get the set of one-dimensional time-dependent differential equations

$$(T_\mathrm{N} + E_j)\chi_j + \sum_k V_{jk}\chi_k = i\,\partial_t\chi_j, \tag{2.23}$$

where $E_j(R)$ is the field-free electronic eigenenergy (potential surface). The radiative coupling, in the length gauge and dipole approximation, is given by

$$V_{jk}(R, t) \equiv \langle\phi_j(\boldsymbol{r}, R)| - \boldsymbol{d}\cdot\boldsymbol{F}(t)|\phi_k(\boldsymbol{r}, R)\rangle, \tag{2.24}$$

where $\boldsymbol{d}$ is the dipole moment. When H_2^+ interacts with a low-frequency field, the 1s σ_g ground state couples strongly to the repulsive 2p σ_u state, but weakly to any other states. Calculations have shown that it is justified to limit the molecular basis to these two states as long as the coupling is expressed in the length gauge [35, 36]. Another useful simplification is to assume that the molecular axis is aligned parallel to the laser field. The resulting equations are trivially solved with modern computing resources; more details can be found in chapter 4. The initial condition $\chi_\mathrm{g}(R, t = 0)$ can be used to define the initial vibrational state of the system.

Figure 2.9 is adapted from the topical review on the dynamics of H_2^+ in intense laser fields [36]. Initially, the $\chi_\mathrm{g}(R, t = 0)$ function describes the $v = 4$ vibrational level. The plot shows the evolution of the nuclear wavepacket, $|\chi_\mathrm{g}|^2 + |\chi_\mathrm{u}|^2$, during a laser pulse with the parameters $T = 150$ fs, $\lambda = 780$ nm and F_0 corresponding to an intensity of 1.7×10^{14} W cm^{-2}. At $t = 0$, we recognise the peaks of the $|\chi_\mathrm{g}(v = 4)|^2$ distribution and in particular the dominance of the peak at $R \approx 3.5a_0$, i.e. the one

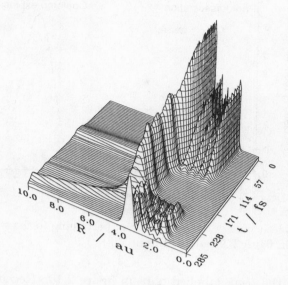

Fig. 2.9. The temporal evolution of a H_2^+ nuclear wave for a pulse of duration $T = 150$ fs, intensity $I = 1.7 \times 10^{14}$ W cm^{-2}, starting from the $v = 4$ level (adapted from [36]).

at the outer turning point of the potential well. From $t \approx 50$–100 fs, we note a steep decline of the bound state peaks while a wavepacket travels outwards, indicating dissociation along the repulsive u-state surface. However, a strongly localised bound state is also formed during this process. This is an example of a laser-induced bound state that evolves on a hybrid potential surface (a dressed state) and under certain conditions becomes trapped. This trapping mechanism is strongly enhanced when there is a degeneracy between the field-free state and the laser-induced state as first noted by Bandrauk and coworkers [37]. As the pulse ends the laser-induced hybrid state reverts to its components and splits into two parts, a dissociating wavepacket and a vibrationally excited bound state. In this case primarily, though not purely, $v = 4$.

The proton-energy spectrum displays two distinct photodissociation channels, as shown in figure 2.10. The first wavepacket, released around $\simeq 100$ fs, has the larger velocity (see figure 2.9) and it contributes only to the structure in the range 0.5–0.8 eV. The wavepacket on the trailing edge of the pulse consists largely of a low-energy part, which is responsible for the distribution around 0.1 eV. However, a close look at figure 2.9 reveals that it also makes a small contribution to the higher-energy peak. The interference is strongly dependent on the delay time, that is pulse duration. Interference with the earlier wavepacket results in the irregular

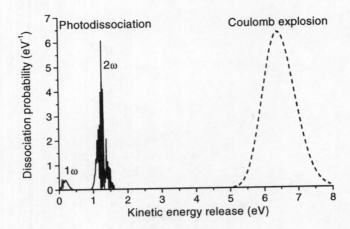

Fig. 2.10.   Kinetic energy release spectrum pertaining to the wavepacket calculation shown in figure 2.9 (adapted from [36]).

structure on the peak labelled '2ω' in figure 2.10. Recent experimental evidence for this trapping effect is being discussed later.

Dissociation in the long-wavelength limit

In the long-wavelength limit, the laser field varies slowly in comparison with nuclear motion. The external field displaces the valence electrons from the nuclear cores, thus weakening the molecular bond [38]. We expect that, near the maxima of the field, dissociative nuclear wavepackets are produced [39]. For short laser periods of the order of 1 fs, heavy nuclei move very little during one laser period. Consequently, the interference of nuclear wavepackets at the position of their creation strongly influences the process of dissociation. In the long-wavelength limit, on the other hand, each wavepacket moves well away from the parent molecule before the next one is produced and the mechanism of dissociation is in principle no different from that of a static field. Still, the interference of the wavepackets in the break-up channel leads to structure in the energy in which the peaks are separated by the photon energy [40]. In analogy to ATI, this process is termed *above-threshold dissociation* (ATD) [41].

We noted in the previous section that an electric field mixes particularly the repulsive 2p σ_u state and the 1s σ_g ground state. The ensuing displacement of electronic charge of the ground state is energetically favourable due to the potential of the external field. The Hamiltonian matrix for the molecule in a static field (F_0) is given by

$$\begin{pmatrix} E_g & V_{gu} \\ V_{ug} & E_u \end{pmatrix},$$

$$(2.25)$$

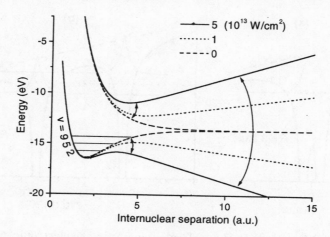

Fig. 2.11. The two lowest electronic states of H_2^+ are strongly mixed by a DC electric field.

where $V_{gu} = V_{ug} = \langle \phi_g | eF_0 z | \phi_u \rangle$. For most purposes, the approximation $V_{ug} \approx \frac{1}{2} F_0 R$ is quite adequate [42]. The corresponding eigenvalues are

$$E'_{g,u} = \frac{1}{2}(E_g + E_u) \pm \frac{1}{2}(E_g - E_u)\left[1 + 4V_{ug}^2/(E_g - E_u)^2\right]^{\frac{1}{2}}. \qquad (2.26)$$

Figure 2.11 illustrates the dependences of E'_g and E'_u on field strength. In the long-wavelength limit, the molecule cycles through these curves. For small internuclear separations, the ground-state electron can adjust adiabatically to the slowly varying field and the state is therefore always described by the lower curve, see figure 2.12(a). The 'clapping' behaviour of the potential curves at *twice* the laser frequency is indicated by the short arrows at $R \approx 5a_0$ in figure 2,11.

For increasing R, a large potential barrier at the centre of the molecule appears, see figure 2.12(b). Since the tunnelling time increases exponentially with the width and height of the barrier, the electron becomes completely localised. As a result, the eigenstates are separated atomic orbitals, i.e. $H^+ + H(1s)$ and $H(1s) + H^+$. On the other hand, the molecular orbitals switch smoothly from ϕ_u to ϕ_g in concert with the field. The long double-headed arrow in the right-hand half of figure 2.11 is intended to illustrate the oscillatory behaviour of the two curves. The wavepacket now moves on a curve that is periodically attractive and repulsive which induces ponderomotive wiggling of the proton.

The localisation is also illustrated by the magnitude of the Stark shift. The g- and u-curves in figure 2.13 are calculated according to $E'_g - E_g$ and $E'_u - E_u$. They show that, for large R, the shifts are indeed equal to $\pm V_{ug}$ (i.e. $\pm \frac{1}{2} FR$), as expected for a localised electron. For small R the

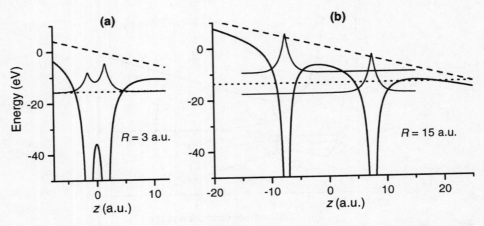

Fig. 2.12.   (a) For small R, the molecule remains in the electronic ground state. The electron cloud oscillates at the same frequency but in antiphase with the laser field. (b) For large R, the two electronic states are localised at the protons. Since the large central barrier impedes tunnelling, the electron remains localised and thus switches periodically between the ground and excited states.

electronic levels split and the shifts are strongly suppressed.

In order to visualise the strong-field dissociation process [40], let us consider a certain vibrational level of the ground state ϕ_g, see figure 2.11. At some stage on the rising edge of the laser pulse, either during a crest or during a trough of the laser electric field, the barrier in the E'_g curve is lowered so that a significant fraction of the *nuclear* wavepacket escapes, either 'over the barrier' or by tunnelling. The wavepacket accelerates on the downward slope beyond the barrier. In the long-wavelength limit, the nuclear wavepacket has moved so that the electron is now completely localised. Consequently, the transition to ϕ_u means that, for the next half

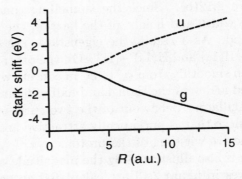

Fig. 2.13.   Stark shifts of the g- and u-states, calculated from equation (2.26) for $F_0 = 0.02$ a.u., which corresponds to $\simeq 1.5 \times 10^{13}$ W cm^{-2}.

of a laser period, the potential is attractive. The motion of the proton is therefore a combination of oscillation and linear translation. Whereas the ponderomotive energy will be appreciable at long wavelengths, the translational or 'drift' energy, particularly if the wavepacket was released exactly at a crest or trough, will be relatively small. This is equivalent to *Simpleman's model* [43], which was used for the interpretation of ATI in the long-wavelength limit. Here too, the drift energy depends on the phase of the laser field at the moment that the wavepacket is released. Since the ponderomotive energy is returned to the laser field on the trailing edge of the pulse, the observed kinetic energy corresponds to the drift energy.

In reality there are several complicating factors. Although a certain degree of alignment of the molecular axis with the laser field is necessary for efficient coupling of the ϕ_g and ϕ_u states, the one-dimensional model is of course an oversimplification. Misalignment results in the proton not wiggling back to its origin, meaning that some energy is gained. Also, due to the uncertainty principle, the wavepacket that is released within a short time window must have a significant energy spread. Finally, part of the wavepacket will not reach large enough R for the electron to be fully localised before the field goes to zero. In this case, the ϕ_g state will make only a partial transition to ϕ_u, i.e. the wavepacket splits into two parts, which evolve independently on the ϕ_g and ϕ_u curves for the next half period. Dietrich *et al.* [44] developed a semiclassical model in which the transitions in the intermediate range of R are described by a Landau–Zener-type formula. Naturally, for shorter wavelengths, the number of transitions increases, until the wavepackets are split and recombined so often that interference effects dominate the dissociation process. Under these conditions a quantal treatment of nuclear motion is essential.

The Floquet theory

When the laser pulses are long and of short wavelengths, the external field is a periodic perturbation $V(\boldsymbol{r}, t + 2\pi/\omega) = V(\boldsymbol{r}, t)$. The molecular dynamics can be understood in terms of the Fourier decomposition of the quantal equations [37, 45, 46]. Consider the electronic dynamics alone for the present. The Floquet theorem states that the solutions to the wave-equation take the form

$$\Psi_j(\boldsymbol{r}, t) = e^{-i\varepsilon_j t/\hbar} \sum_{N=-\infty}^{+\infty} e^{-iN\omega t} \Phi(j, N; \boldsymbol{r}), \qquad (2.27)$$

where the quasi-energy, ε, is in general complex. It is clear that ε is indeterminate to within multiples of $\hbar\omega$, i.e. identical solutions exist for $\varepsilon \pm \hbar\omega$, $\varepsilon \pm 2\hbar\omega$, etc. In the absence of molecule–field interaction, the Floquet wavefunctions, Φ, are the usual eigenstates of the unperturbed molecule, which are now 'dressed' with the phase factors $e^{-iN\omega t}$. With interaction, however, the Floquet Ansatz (2.27) converts the time-dependent Schrödinger equation into an infinite set of time-independent differential equations, in which neighbouring Fourier components are coupled. The mixing of adjacent dressed states with phase factors $e^{-iN\omega t}$ and $e^{-i(N\pm 1)\omega t}$ leads to oscillations of the electron at the driving frequency ω; an effect enhanced if $\hbar\omega$ is close to the difference in energy between two states. Non-neighbouring dressed states are indirectly coupled, via the neighbours in between. Thus higher-order harmonic oscillations also occur, though with progressively smaller amplitudes and stricter resonance conditions.

It is possible to interpret the Floquet dressed states in terms of photons of the quantized radiation field. The total Hamiltonian consists of three parts, i.e. the molecule, the field and the molecule–field interaction term. In the system as a whole energy is conserved. Energy lost by the field is absorbed by the molecule and vice versa. The laser field can be represented in terms of Fock states: occupation numbers of the mode of a quantised field of bosons, $|n\rangle$ [47]. In the limit $n \gg 1$, corresponding to intense lasers and large numbers of photons, the quantised field is equivalent to the semiclassical method. In particular the Fourier harmonic of order N corresponds to the occupation of the field-mode by N photons. Figure 2.14(a) shows the potential curves of the g- and u-states dressed with N, $N \pm 1$, $N \pm 2$, ... photons of 1.5 eV and (b) displays the avoided crossings resulting from radiative couplings.

Adiabatic potential curves

The laser field has the time dependence $F_0 \cos(\omega t)$ so that the perturbing potential can be written as

$$V(\boldsymbol{r}, t) = -\boldsymbol{d} \cdot \boldsymbol{F}(t) = V_- e^{i\omega t} + V_+ e^{-i\omega t}. \qquad (2.28)$$

The Fourier components then obey the infinite set of time-independent differential equations

$$(\varepsilon_j + N\hbar\omega - H_e)|\Phi(j, N)\rangle = V_+|\Phi(j, N-1)\rangle + V_-|\Phi(j, N+1)\rangle. \qquad (2.29)$$

The interpretation of $\Phi(j, N)\rangle$ as a state dressed with N photons indicates that V_- is responsible for one-photon emission and, conversely, V_+ is responsible for absorption. Parity selection rules apply, so that only g- and u-states are mixed by this operator. If we restrict the electronic set of states to the two lowest states only (1s σ_g, 2p σ_u), the resulting set of

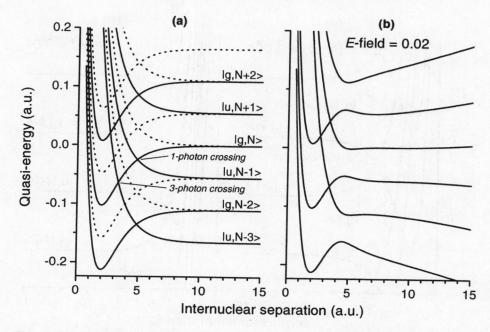

Fig. 2.14.   (a) The photon-dressed states of H_2^+ without interaction are called diabatic curves. The labels, $|g,N>$ etc., belong to the solid curves. (b) With a strong interaction (here $F_0 = 0.02$ a.u.) the periodicity of the quasi-energies is perturbed due to truncation of the size of the Floquet matrix.

equations for the quasi-energy is given by the eigenvalues of the matrix

$$
\begin{pmatrix}
\cdots & \cdots & \cdots & \cdots & \cdots & \cdots \\
\cdots & D_u(N+1) & V_+ & 0 & 0 & \cdots \\
\cdots & V_- & D_g(N) & V_+ & 0 & \cdots \\
\cdots & 0 & V_- & D_u(N-1) & V_+ & \cdots \\
\cdots & 0 & 0 & V_- & D_g(N-2) & \cdots \\
\cdots & \cdots & \cdots & \cdots & \cdots & \cdots
\end{pmatrix}, \quad (2.30)
$$

where N runs from $-\infty$ to $+\infty$ and the diagonal terms are $D_g(N) = N\hbar\omega - E_g$, $D_u(N-1) = (N-1)\hbar\omega - E_u$. The off-diagonal coupling terms are $V_\pm = \hbar\Omega_R/2 \equiv (F_0/2)\langle 2p\,\sigma_u|d|1s\,\sigma_g\rangle$, where Ω_R is the Rabi frequency.

The infinite Floquet matrix must be truncated in practical calculations. The size of the matrix must be such that the periodicity (ϵ_j, $\epsilon_j \pm 2\hbar\omega$, ...) is accurately reproduced and the results are insensitive to further increments of the size of the matrix. An example of the Floquet energies for $F_0 = 0.02$ a.u. $(1.4 \times 10^{13}$ W cm$^{-2})$ is shown in figure 2.14(b). The quasi-energies at $F_0 = 0.004$, shown in figure 2.15(a), are the central

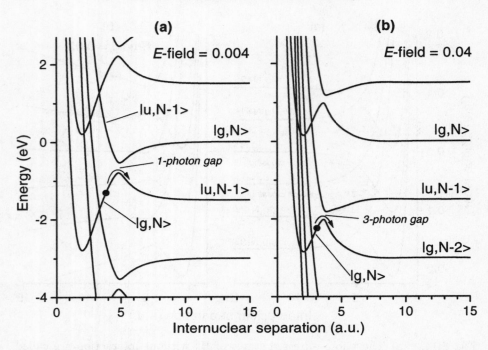

Fig. 2.15.   (a) At $F_0 = 0.004$ a.u. $(6 \times 10^{11}$ W cm$^{-2})$, the adiabatic dressed states are still similar to the diabatic states of figure 2.14, except for the small one-photon gap. (b) At $F_0 = 0.04$ a.u. $(6 \times 10^{13}$ W cm$^{-2})$, a gap has opened at the three-photon crossing. The one-photon gap is now so large that it is no longer recognisable.

eigenvalues of a 10×10 matrix and those at $F_0 = 0.04$ are those of an 18×18 matrix, see figure 2.15(b).

Following conventions established in the field of atomic collisions, the unperturbed curves in figure 2.14(a) are called *diabatic*, whereas those of figure 2.15(a) are labelled *adiabatic*. We note that the adiabatic curves at the lower field of $F_0 = 0.004$ are fairly similar to the diabatic ones, except that the one-photon crossing is replaced by an anti-crossing. The interaction is strong at the crossing where the transition $|g\rangle \rightarrow |u\rangle$ is resonant. In this region the electronic states are mixed. At the crossing point the Floquet states change from pure $|g, N\rangle$ and $|u, N - 1\rangle$ to $2^{-1/2} (|g, N\rangle \pm |u, N - 1\rangle)$. Physically the entire electronic charge oscillates between the two protons, in phase with the laser field.

Bond-softening and above-threshold dissociation

Suppose that the protons of H$_2^+$ could be described by classical mechanics and thus could have a definite separation between them. When R is fixed and the laser intensity is increased slowly, the diabatic curves which describe the electronic states evolve into the adiabatic ones. Alternatively we may keep the field constant and slowly vary R. Consider the $|g, N\rangle$ ground state dressed with N photons at $F_0 = 0.004$ and with R to the left of the one-photon gap, as indicated in figure 2.15(a) by the black dot. If the protons now slowly move apart, the electron–field state will follow the adiabatic curve, thus following the lower branch of the adiabatic crossing shown in figure 2.15(a). The molecule is gradually going from the ground state to the repulsive state whilst the field is losing one photon. This process of adiabatic dissociation of H_2^+ by the absorption of one photon is conventional photodissociation. However, when the field is extremely intense, the large distortion of the potentials means that perturbation theory no longer applies. This effect is usually called *bond-softening* [10, 48] as it tends to lower the potential barrier and to weaken the restoring force in the well.

At the higher field of $F_0 = 0.04$, a gap has opened at the three-photon crossing. The one-photon gap is now so large that it is no longer recognisable, since $\Omega_R > \omega$. (In fact, at such intensities higher excited states and the ionisation continuum should be included.) Following the lower branch of the three-photon gap (see the arrow in figure 2.15(b)), the $|g, N\rangle$ makes an adiabatic transition to the $|g, N-2\rangle$ state. This is normally interpreted as three-photon absorption on the lower branch of the three-photon gap, followed by the emission of one photon on the upper branch of the one-photon gap between the $|g, N-2\rangle$ and $|u, N-3\rangle$ states. However, the three-photon gap is not simply a periodic mixing of $|g, N\rangle$ and $|u, N-3\rangle$ at a frequency of 3ω; there will also be oscillations at the fundamental frequency. Consequently, the three-photon gap is lowered with respect to the diabatic three-photon crossing (cf. figures 2.14(a) and 2.15(b)).

When two photons are absorbed, even though one photon would have sufficed energetically, the dissociation process is usually termed *above-threshold dissociation* (ATD) [41]. Conventional conservation of energy dictates that dissociation with the absorption of one photon is impossible when the molecular binding energy is larger than the photon energy. In this case, the dissociation via the three-photon gap might be termed two-photon bond-softening, rather than ATD. When the H_2^+ molecule is prepared in a non-resonant way, e.g. via tunnel ionisation, a certain range of vibrational levels is populated. Consequently, one- and two-photon bond-softening yields two peaks in the photodissociation spectrum whose releases of kinetic energy are separated by a little less than $\hbar\omega$. Indeed, a third peak would imply the occurrence of ATD, as long as the Coulomb-explosion channel, $H^+ + H^+$, can be excluded as its cause [10]. A more detailed test of ATD will require an experiment in which the H_2^+ ions are prepared in pure vibrational states via a resonance process at short wave-

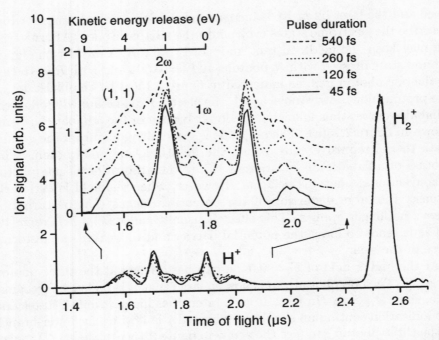

Fig. 2.16. Ion time-of-flight spectra of H_2 subjected to ultra-short intense laser pulses at $\lambda = 792$ nm, peak intensities of 1.5×10^{14} W cm^{-2} and four different pulse lengths. Adapted from [50].

length [49], before they are subjected to intense pulses in the infrared region.

Below-threshold dissociation and zero-photon dissociation

When pulses are short the Floquet approach must be modified to take into account the change in field-amplitude by noting that the curves are now time dependent and move according to the variation in laser intensity. We review an experimental study in which H_2 molecules are being subjected to intense ultra-short pulses whose duration is varied from 45 to 540 fs, by frequency chirping the pulses of an intense Ti : sapphire laser beam [50], and hence is able to probe the dynamic potential surfaces.

Figure 2.16 shows ion time-of-flight spectra recorded at four different pulse lengths. In each case the peak intensity was adjusted to about 1.5×10^{14} W cm^{-2}, i.e. below saturation of any ion channel [15] (cf. figure 2.8), and such that the H_2^+ peak in each spectrum was of the same height. Comparing these spectra with the theoretical results presented in figure 2.10 suggests that the two lower-energy channels, labelled 1ω and 2ω, are due to photodissociation of H_2^+ from the $v = 4$ vibrational level.

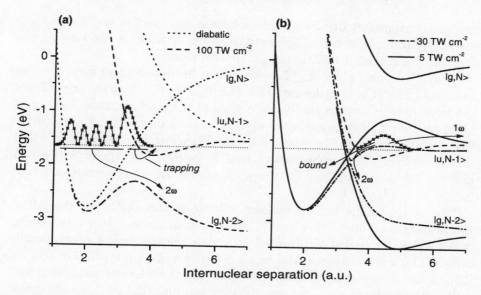

Fig. 2.17. Adiabatic curves showing the routes towards the two-photon bond-softening and below-threshold dissociation of H_2^+ ($v = 4$) in ultra-short, intense laser pulses at $\lambda = 792$ nm.

The covariance-mapping method [16], which is described in chapter 3, indeed confirmed that the higher-energy channel, labelled (1,1), originates from the Coulomb explosion (CE) of two protons. The CE peak shifts to higher kinetic energies with pulse shortening, because a shorter rise time initiates Coulomb explosion at a smaller internuclear distance, as discussed later in this chapter.

The binding energy of the $v = 4$ vibrational level is larger than the photon energy. The dissociation of H_2^+ on the leading edge of the pulse starts, therefore, when the three-photon gap opens, see figure 2.17(a). As a result, a wavepacket travels to the $|g, N - 2\rangle$ curve by absorbing two photons in total (see also figure 2.15(b)).

However, not all of the population follows this path; at the outer turning point of the vibrational level, the initial wavepacket is in a good position to start moving on the upper branch of the adiabatic curves. We note that the anti-crossing induces a potential well above the three-photon gap around $R = 4a_0$. The stable wavepacket of figure 2.10 that survives the height of the laser pulse from $t \approx 100$–200 fs is trapped in this well [51, 52].

On the trailing edge of the laser pulse, the laser-induced potential well dissolves in a peculiar manner, see figure 2.17(b). On the near side, it is transformed into the $|u, N - 3\rangle$ and the $|g, N\rangle$ diabatic dressed states as one might expect for a three-photon gap. Thus around $R \simeq 4a_0$, the trapped wavepacket breaks up into a bound wavepacket and a dissociating wavepacket, which still follows the upper branch at the one-photon crossing, such that it contributes to the 2ω peak.

On the far side, i.e. for $R \geq 5a_0$, the upper branch changes from concave to convex and it is transformed into the $|u, N - 1\rangle$ diabatic curve. Since the laser-induced well is shallow and wide at this end, some population is lifted by the quickly rising curve and dissociates on the $|u, N - 1\rangle$ curve. Clearly, the faster the intensity falls the higher the wavepacket is lifted and the larger the kinetic energy that it gains. This dynamic Raman process [50] is responsible for the shift of the 1ω peak shown in figure 2.16.

The escape of the 1ω wavepacket is quite remarkable since the initial vibrational level lies below the $|u, N - 1\rangle$ continuum. In other words, dissociation by the absorption of only one photon seems energetically impossible, which suggests the name *below-threshold dissociation* (BTD) [53]. This conceptual problem arises because the Floquet method is not well suited for ultra-short pulses. Keeping in mind that photons are being absorbed and re-emitted every laser period, we note that the symmetry between absorption and emission is broken by the quickly changing laser intensity. Large energy gains are possible by multiple absorption from the high-energy end of the spectrum (45 meV FWHM) and re-emission at the low-energy end. The final $|N - 1\rangle$-photon state merely testifies to the *nett* absorption of one photon.

This experiment seems to substantiate the theoretical predictions on vibrational trapping above the three-photon gap. Vibrational trapping above the one-photon crossing is of course also possible, but generally requires higher initial vibrational levels and shorter wavelengths. Figure 2.18, which was adapted from [51], illustrates this for the $v = 10$ vibrational level and $\lambda = 266$ nm. At around 10^{13} W cm^{-2}, the one-photon gap opens up and a dissociating wavepacket can travel towards the $|u, N - 1\rangle$ curve, resulting in the 1ω dissociation channel. At the same time a large wavepacket is trapped in the laser-induced well which is formed above the crossing.

We noted earlier that vibrational trapping above the three-photon gap can lead to the formation of a slowly dissociating wavepacket on the trailing edge of the laser pulses. With the one-photon well, a similar pulse-shape effect is possible on the leading edge of the pulse. If the laser intensity is allowed to increase further, the well becomes shallower and the wavepacket can partially escape to large R before the intensity drops again on the trailing edge of the pulse. Although the potential well does not disappear completely until 4×10^{14} W cm^{-2}, population can escape well before that, because the trapped wavepacket is vibrationally excited. Since the wavepacket dissociates on the $|g, N\rangle$ curve, no nett number of

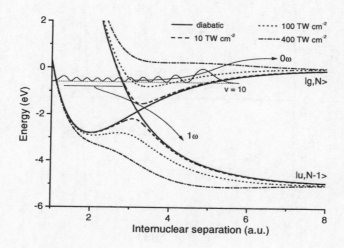

Fig. 2.18. Adiabatic curves illustrating the routes towards the bond-softening (1ω) and zero-photon dissociation (0ω) of H_2^+ ($v = 10$) in ultra-short, intense laser pulse at $\lambda = 266$ nm (based on [51]).

photons is absorbed and one might therefore call this channel zero-photon dissociation (ZPD).

In another recent experiment, H_2 was being subjected to intense ultra-short pulses at $\lambda = 400$ nm and $\lambda = 266$ nm of a frequency-doubled and -tripled Ti : sapphire laser beam [54]. Two representative time-of-flight spectra are presented in figure 2.19. Besides the ordinary 1ω bond-softening peaks, these spectra also display very-low-energy protons, which are ascribed to ZPD.

We noted already that a more detailed investigation of ATD would require the preparation of H_2^+ in a well-defined vibrational state. It will be clear from the previous discussions that this is equally true for BTD and ZPD, which are also very sensitive to the initial vibrational state. However, the ionisation of H_2 by ultra-short, intense laser pulses is still poorly understood.

2.5 Strong-field ionisation and Coulomb explosion

We discussed in section 2.3 how strong-field ionisation is divided into three regimes, according to the value of the Keldysh parameter, namely multi-photon, tunnelling and over-the-barrier ionisation. In this chapter we concisely treat the Floquet method, the quasi-static field ionisation method and a classical over-the-barrier model that pertain to these

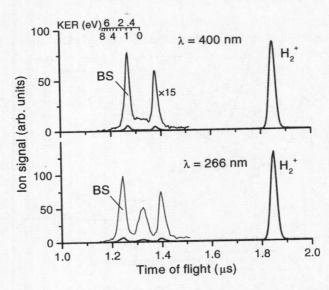

Fig. 2.19. Ion time-of-flight spectra of H_2 subjected to frequency-doubled (400 nm) and -tripled (266 nm) pulses of an ultra-short, intense Ti : sapphire laser (adapted from [54]).

regimes, as well as the general wavepacket method.

Perturbation theory

At very high intensity the effect of the external field dominates the electronic dynamics. Under such conditions one can visualise the process as the electrons being suddenly stripped from the parent molecule. In such circumstances the electronic potential well (and hence the energy spectrum) is severely distorted by the external field to such an extent that the excited states are very weakly bound and have large energy shifts and the ejection mechanism is similar to field ionisation. In summary, the molecular potential does not play an important role in the ionisation process itself, but merely serves to provide the initial conditions, namely the binding energy and momentum distribution for the electron. Of course, these initial conditions are vital in determining the rate of ionisation. In collision physics this approximation is analogous to the *impulse approximation*. Given that the binding potential of the parent molecule can be considered weak with respect to the external forces, it is possible to develop a perturbation series by expansion in terms of this potential. In the approximation introduced by Keldysh [29] and subsequently refined [55, 56] and extended by others [31, 57], the effect of the residual molecular ion on the photoelectron is completely neglected, but the interaction of the photoelectron with the external field is taken into account to all orders. In the same spirit, the initial state is assumed to be an unperturbed molecular orbital; the binding potential is treated exactly but the external field is neglected. Let us consider the formulation of this model for a simple one-electron molecular system. The electronic motion is governed by the local potential $U(\boldsymbol{r})$ and the external electric field $\boldsymbol{F}(t) = \boldsymbol{F}_0 \cos(\omega t)$. Then, within the dipole approximation, the Hamiltonian takes the form

$$H(t) = \tfrac{1}{2}\boldsymbol{p}^2 + \boldsymbol{A}(t){\cdot}\boldsymbol{p} + \tfrac{1}{2}\boldsymbol{A}^2(t) + U(\boldsymbol{r}) + W(R), \qquad (2.31)$$

in which $\boldsymbol{p}$ is the momentum operator. We have used atomic units and defined the vector potential in the usual way: $\boldsymbol{A}(t) = -\int^t \boldsymbol{F}(t')\,\mathrm{d}t'$. Within the dipole approximation, the quadratic term $\tfrac{1}{2}\boldsymbol{A}^2(t)$ and the internuclear potential $W(R)$ do not give rise to electronic transitions and can be removed by a simple unitary transformation:

$$H'(t) - \mathrm{i}\partial_t = U^{\dagger}(H - \mathrm{i}\partial_t)U, \qquad (2.32)$$

where

$$U = \exp\!\left(-\mathrm{i}\int^t \left[\tfrac{1}{2}\boldsymbol{A}^2(t') + W(R)\right]\mathrm{d}t'\right), \qquad (2.33)$$

in which

$$H'(t) = \tfrac{1}{2}\boldsymbol{p}^2 + \boldsymbol{A}(t){\cdot}\boldsymbol{p} + U(\boldsymbol{r}). \qquad (2.34)$$

The molecular orbital (initial state), unperturbed by the external field, satisfies the equation

$$\left[\tfrac{1}{2}\boldsymbol{p}^2 + U(\boldsymbol{r})\right]\phi_i = E_i\phi_i, \qquad (2.35)$$

where E_i is the electronic energy for a given value of R. The photoelectron is assumed to be unaffected by the residual ion and satisfies the equation

$$\left(\tfrac{1}{2}\boldsymbol{p}^2 + \boldsymbol{A}(t){\cdot}\boldsymbol{p} - \mathrm{i}\partial_t\right)\chi_k = 0, \qquad (2.36)$$

where the solutions are termed Volkov waves, which take the form

$$\chi_k \equiv (2\pi)^{-3/2}\exp(\mathrm{i}\boldsymbol{k}{\cdot}\boldsymbol{r} - \mathrm{i}\boldsymbol{k}{\cdot}\boldsymbol{\alpha}_0\cos(\omega t) - \mathrm{i}\tfrac{1}{2}k^2 t), \qquad (2.37)$$

in which $\boldsymbol{\alpha}_0 = \boldsymbol{F}_0/\omega^2$. The Volkov wave is not an eigenfunction of the time-indepedent Hamiltonian since it includes all harmonics/orders of the field. This is clear from the Fourier decomposition,

$$\exp[-\mathrm{i}\boldsymbol{k}{\cdot}\boldsymbol{\alpha}_0\cos(\omega t)] = \sum_{N=-\infty}^{+\infty} \mathrm{i}^N J_N(\boldsymbol{k}{\cdot}\boldsymbol{\alpha}_0)\exp(\mathrm{i}N\omega t), \qquad (2.38)$$

where J_N is a Bessel function. In principle an infinite range of photoelectron energies can be populated; however, the weighting of this spectrum is strongly dependent on the momentum distribution of the initial state. For example photo-ionisation following N-photon absorption will lead to an electron with energy $E_{fN} = \frac{1}{2}k_N^2$ with

$$\frac{1}{2}k_N^2 = E_i + \Delta_i + N\omega, \tag{2.39}$$

where we have adjusted the binding energy in order to allow for the effect of the Stark shift acting on the initial state. To a good approximation, for the ground state of the molecule and for low-frequency light, $\Delta_i \approx -U_{\mathrm{p}} = -F_0^2/(4\omega^2)$. This adjustment greatly improves agreement with more sophisticated models [58]. Applying time-dependent perturbation theory leads to the Fermi golden rule for the photo-ionisation of the molecule by N photons. Below threshold ($k_N^2 < 0$) the rate vanishes: $\Gamma_N = 0$. However, above threshold

$$\Gamma_N = 2\pi k_N (N\omega + \Delta_i)^2 \int d\hat{\boldsymbol{k}}_N J_N^2(\boldsymbol{k}_N \cdot \boldsymbol{\alpha}_0) |\tilde{\phi}_i(\boldsymbol{k}_N)|^2, \tag{2.40}$$

for $k_N^2 > 0$, where the wavefunction in momentum space is

$$\tilde{\phi}_i(\boldsymbol{k}) \equiv (2\pi)^{-3/2} \int d\boldsymbol{r} \exp(-i\boldsymbol{k}\cdot\boldsymbol{r}) \, \phi_i(\boldsymbol{r}). \tag{2.41}$$

In the weak-field limit, the Bessel function can be expanded as $J_N(z) \approx z^N$ and thus $\Gamma_N \propto I^N$, as expected from leading-order-perturbation approximation. The total rate of ionisation is given by

$$\Gamma = \sum_N \Gamma_N. \tag{2.42}$$

This model allows us to estimate the rate of ionisation by simple computations. The only element of quantum chemistry is the determination of the Fourier transform of the bound-state function and this is a straightforward task. For example, if we consider the ground state of H_2^+ and take the simple linear combination of atomic orbitals [59] with an effective charge Z, we obtain

$$\tilde{\phi}_i(\boldsymbol{k}) = \frac{4}{\pi\sqrt{1+s}} \frac{\cos(\frac{1}{2}\boldsymbol{k}\cdot\boldsymbol{R})}{(1+k^2)^2}, \tag{2.43}$$

where the atomic-orbital-overlap integral is $s = (1 + ZR + \frac{1}{3}Z^2R^2)e^{-ZR}$. As in the simple one-photon ionisation process, in the dipole approximation, the photon field does not provide momentum to the photoelectron. This is supplied from the bound-state velocity distribution and explains the shape of the photoelectron-energy spectrum. This spectrum is also very sensitive to the laser polarisation, expressed by the factor

$J_N^2(\boldsymbol{k}_N \cdot \boldsymbol{\alpha}_0)$, when typically $N \gg 1$. The effect of Bragg scattering is represented by the factor $\cos(\frac{1}{2}\boldsymbol{k} \cdot \boldsymbol{R})$. In physical terms this represents the scattering of the electron between the nuclei and is familiar from the context of scattering of fast electrons from molecules. The Keldysh model can readily be extended to treat many-electron targets and also photodissociative ionisation within the Franck–Condon approximation.

Wavepacket methods

Many theoretical simulations have been applied to the model problem of multi-photon ionisation of the molecular hydrogen ion. In the simplest calculation the nuclei are fixed in space and the electronic dynamics are determined. At low frequencies the process is dominated by the large dipole moment along the molecular axis giving rise to alignment [60]. Essentially this is a consequence of the large energy gap between the ground and first excited states so that, for low frequencies, the most strongly off-resonant processes are driven by parallel transitions. If the laser is linearly polarised along the axis, the electronic wavefunction has cylindrical symmetry and the evolution occurs in only $(2+1)$ dimensions. The full panoply of methods for solving partial differential equations can be brought to bear: finite differences, spectral methods, Galerkin methods, finite elements, discrete-variable representations and so forth. If the light is extremely intense and of short duration with a finite bandwidth, then the direct solution of the temporal evolution is essential, with the advantage of providing a visualisation of the physics of ionisation. Furthermore, wavepacket methods are readily applied to describe processes in the energy domain (long pulses of low intensity) as well as processes in the time domain (short pulses of high intensity).

Consider a model problem in which the molecular ion is at the equilibrium internuclear separation ($R = 2a_0$) and irradiated by a pulse of light corresponding to an angular frequency of $\omega \approx 0.2$ a.u., which is equivalent to a wavelength of $\lambda = 228$ nm. We can solve this problem in cylindrical coordinates, with ρ the radial distance and z the axial coordinate, by finite differencing of the time-dependent equation:

$$H_e(R; \rho, z, t)\psi(R; \rho, z, t) = i\hbar \frac{\partial}{\partial t}\psi(R; \rho, z, t). \qquad (2.44)$$

The wavefunction $\psi(R; \rho, z, t)$ is defined on a uniform grid with spacings $\Delta\rho$ and Δz; then, given $\psi(R; \rho, z, t)$, the aim is to determine $\psi(R; \rho, z, t + \delta t)$. An *explicit* integration scheme is the Euler integration:

$$\psi(t + \delta t) \approx \left(1 - \frac{i\,\delta t}{\hbar} H_e\right)\psi(t), \qquad (2.45)$$

while a simple *semi-implicit* method is use of the Crank–Nicholson formula

$$\left(1 + \frac{\mathrm{i}\,\delta t}{2\hbar} H_{\mathrm{e}}\right)\psi(t + \delta t) = \left(1 - \frac{\mathrm{i}\,\delta t}{2\hbar} H_{\mathrm{e}}\right)\psi(t). \tag{2.46}$$

Implicit propagators, while possessing the quality of stability, are computationally expensive and rather inefficient. Explicit schemes can be highly accurate and efficient if the time step is chosen sufficiently small. The Taylor series is an explicit single-step propagator. For short time steps, such that $|(\partial H_{\mathrm{e}}/\partial t)\,\delta t| \ll |H_{\mathrm{e}}|$, the Nth-order Taylor series for $\psi(t + \delta t)$ is

$$\psi(t + \delta t) = \sum_{k=0}^{N} \left(\frac{-\mathrm{i}\,\delta t}{\hbar}\right)^{k} \left(\frac{H_{\mathrm{e}}^{k}}{k!}\right)\psi(t). \tag{2.47}$$

In Figure 2.20 ρ–z plots of the electron probability density of H_2^+ at $R = 2.0a_0$ for $\lambda = 228$ nm and a peak intensity $I = 6 \times 10^{14}$ W cm^{-2} are presented [61]. Figure 2.20(a) corresponds to the unperturbed ground state of the molecular ion. In Figure 2.20(b), corresponding to an instant during the pulse rise at which the field passes through a maximum in the positive z-direction, the density is very slightly distorted along the negative z-axis. The wavepacket behaves as a tightly bound forced oscillator responding in antiphase to the forcing frequency [62]. In Figure 2.20(c), half a cycle later, the bound probability density stretches along the positive z-axis. In Figure 2.20(d) the wavepacket has now a more complex structure as it begins to break up. This can be attributed to the recollision of the ionising wavepacket (apparent in Figure 2.20(c)) with the nuclei and the parent (bound) wavepacket. Later, see figure 2.20(e), the probability density has a larger radial extent and, being loosely held, responds in phase with the field [62, 63].

The rate of ionisation (Γ) is defined as the rate of decay of population in a volume surrounding the nuclei. In figure 2.21 the results for a fourth-order ($N = 4$) series for $\delta t = 0.01$ and $\delta t = 0.001$ are shown. These results concur with higher-order ($N = 10$) calculations. The rate of decay is slow and steady, indicating that the rate of ionisation is well defined for this wavelength, bond length and intensity.

A robust test of the accuracy of wavepacket methods is the comparison of the rates of photo-ionisation with highly accurate Floquet results [64, 65]. The Floquet method [64] gives an extensive coverage of the Hilbert space and, in particular, the spectrum of low-lying excited states is well reproduced. These states are often brought into play by resonant enhancement of the rate of ionisation, leading to a complex frequency and intensity dependence.

Our choice of laser parameters, see table 2.1, characterises a transition in the multi-photon regime that is reasonably well described by pertur-

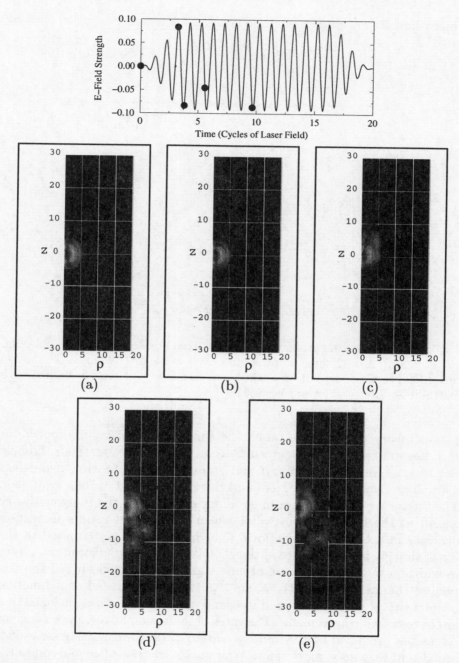

Fig. 2.20. ρ–z plots of the electron probability density of H_2^+ at $R = 2a_0$ for a laser pulse at $\lambda = 228$ nm and a peak intensity of $I = 3 \times 10^{14}$ W cm^{-2}. The picture at the top shows a profile of the laser pulse with the circles corresponding (from left to right) to the location of frames (a)–(e).

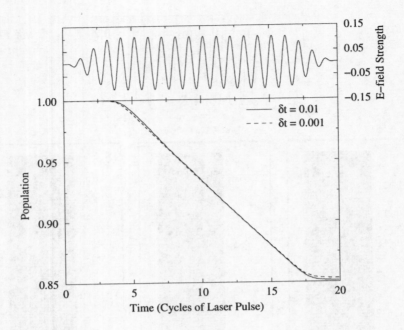

Fig. 2.21. The electron probability within a fixed box around the H_2^+ molecular ion ($R = 2a_0$); $\lambda = 228$ nm and peak intensity $I = 4 \times 10^{14}$ W cm^{-2}. The curves correspond to time steps of $\delta t = 0.001$ a.u. and $\delta t = 0.01$ a.u. In both cases a fourth-order Taylor series was used [61].

bation theory [66]. In the table, the time-dependent results for H_2^+ at $R = 2a_0$ are compared with results from several other methods, namely the Floquet results of Madsen and Plummer [67], the time-dependent method of Chelkowski *et al.* [68] and the Padé method of Baik *et al.* [66]. The Taylor-series results are in good agreement with the Floquet theory except at the highest intensity for which the Floquet results indicate a decrease in the rate of ionisation. This drop in rate is attributed to the Stark shifting into resonance of the $\Sigma_u 2p$ and $\Sigma_u 3p$ five-photon resonance in addition to the suppression of ionisation due to the closing of the six-photon channel at $I \approx 10^{15}$ W cm^{-2}. The general trend as a function of intensity is very well reproduced and the absolute rates of ionisation are in reasonable agreement. The rates of ionisation at the lower range of intensities predicted by the Padé extension of the perturbation series [66] are also in very good agreement. Wavepacket methods become computationally inefficient for large α_0, that is low frequencies, since a large box is required to treat the escaping electron and the highly-excited bound states [61].

In Figure 2.22 a comparison of time-dependent and time-independent

Table 2.1. *A comparison of the rates of ionisation (in s^{-1}) for H_2^+ at $R = 2a_0$ [61] (using grid spacings $\Delta\rho = 0.10a_0$ and $\Delta z = 0.20a_0$) with the Floquet calculation of Madsen and Plummer [67], the Bessel-basis time-dependent calculation of Chelkowski et al. [68] and the perturbative calculation of Baik et al. [66]; a laser pulse of wavelength 228 nm (equivalent to a photon energy of $\omega = 0.2$ a.u.) is used*

Method	Laser intensity (10^{14} W cm^{-2})					
	1.0	2.0	3.0	3.9	5.0	6.0
Taylor series	1.1×10^{11}	1.8×10^{12}	7.7×10^{12}	1.6×10^{13}	3.8×10^{13}	6.6×10^{13}
Floquet	1.4×10^{11}	1.6×10^{12}	5.5×10^{12}	1.4×10^{13}	3.4×10^{13}	2.5×10^{13}
Chelkowski			3.0×10^{12}	1.0×10^{13}	5.0×10^{13}	2.0×10^{13}
Baik	2.1×10^{11}	1.0×10^{12}	5.7×10^{12}	1.6×10^{13}	4.5×10^{13}	9.5×10^{13}

methods is given for $\lambda = 248$ nm, $I = 10^{14}$ W cm^{-2} and a range of internuclear separations. The Keldysh parameter varies between $\gamma \approx 5$ for $R = 2a_0$ and $\gamma \approx 3$ for large R, defining the process as a multi-photon detachment rather than tunnelling ionisation. Very good agreement is found at small internuclear distances at which the ground electronic state is an isolated resonance of the system requiring six photons to reach the ionisation threshold. At larger R, the process requires only four photons to achieve ionisation. The Floquet rates of ionisation refer to a particular dressed state formed from the 1s σ_g ground state. However, around intermediate values of R, the rate of ionisation of the ground state is difficult to define. As a result of crossings with excited states and Rydberg-series resonances, the ground state becomes mixed (dressed) with other states. This mixing is strongly dependent on the pulse risetime. This shows the difficulty in defining a rate of ionisation and hence making a comparison with Floquet theory for interfering resonance states. The results for $R \to \infty$ converge to a separated-atom-ionisation rate of $\Gamma = 1.44 \times 10^{-3}$ a.u., which is in very good agreement with the rate of ionisation of atomic hydrogen.

The Floquet method can readily be extended to multi-electron targets by employing electron–molecule collision theory. The R-matrix Floquet method developed by Burke and coworkers [69, 70] is able to describe two-electron excitation processes including autoionising states in an accurate manner. In the perturbation regime, B-spline techniques have proved extremely powerful in predicting the ionisation rate [71].

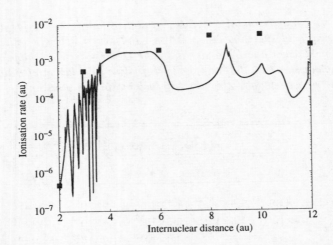

Fig. 2.22.   A comparison of the wavepacket results (squares) [61] for the rate of ionisation as a function of internuclear separation for $\lambda = 248$ nm and $I = 10^{14}$ W cm^{-2} with the Floquet results (full line) of Madsen and Plummer [67].

Over-the-barrier ionisation of diatomic molecules

If a molecule were instantaneously subjected to a field of one atomic unit, undoubtedly it would be field-ionised and lose its valence electrons in less than a femtosecond. Without these electrons to form a bond, the multiply charged molecule would immediately thereafter dissociate on repulsive potential curves that are purely Coulombic for fully stripped molecules. A q-times-ionised molecule will break up into two fragments of charge q_1 and q_2, such that $q_1 + q_2 = q$. For pure Coulomb explosion from the molecule's equilibrium distance, R_e, the kinetic energy of both fragments together would be (in atomic units)

$$E_e = \frac{q_1 q_2}{R_e}. \tag{2.48}$$

The complete fragmentation of H_2 into two electrons and two protons is readily observed with femtosecond laser pulses at intensities of the order of 10^{14} W cm^{-2}. Complete fragmentation of heavier molecules such as N_2 and I_2 requires extremely high intensities, but channels as high as $N^{5+} + N^{5+}$ and $I^{7+} + I^{7+}$ can be observed at intensities around 10^{16} W cm^{-2}. However, the measured release of kinetic energy is generally considerably less than E_e, even in the case of doubly ionised hydrogen, which is definitely Coulombic (the repulsive curves of channels such as $N^{2+} + N^{2+}$ can deviate considerably from Coulombic for short R [72]).

The reason for this discrepancy is of course that the laser pulses have a finite risetime. Rather than being multiply ionised instantaneously, the

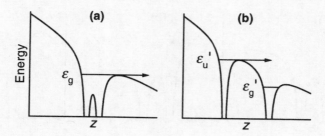

Fig. 2.23. (a) For small R, over-the-barrier ionisation occurs from an almost unshifted ground state. (b) For large R, the threshold is lowest for the upper state. Half a laser period later, the lower state has become the upper state and is ionised.

molecule is sequentially ionised with nuclear dynamics occurring between the steps [73]. In the case of H_2, for example, it is believed that the first ionisation step may well occur at the equilibrium distance, or close to it, but that the process of Coulomb explosion is preceded by a significant amount of bond-lengthening due to the photodissociation of H_2^+. One would expect that pulses with a longer risetime would leave more time between the steps and lead to a larger reduction of the release of kinetic energy.

The simple over-the-barrier model reproduces the threshold intensities for field ionisation of *atoms* well enough [74, 75] to justify its use for the understanding of basic principles. The model assumes that the probability of ionisation jumps from zero to unity as soon as the potential barrier, which is formed by the ionic core and the laser electric field, drops below the energy level of the outer electron. The Stark shift that is induced by the field is neglected; its inclusion would lead to an insignificant change in the threshold intensity for the ionisation of atoms. We noted already (see figure 2.13) that the Stark shift in molecules is also very small as long as the atomic orbitals have a considerable overlap. So, for small R, we can simply apply the atomic model to the molecule in the ϕ_g state and find the minimum electric field that suppresses the barrier below the ϵ_g level, as illustrated in figure 2.23(a). For large R, on the other hand, the molecule is as likely to be in the ϕ_u state as in the ϕ_g state, which are both shifted by a large amount, i.e. $\Delta\epsilon \approx \pm\frac{1}{2}RF$. It is clear from figure 2.23(b) that the threshold intensity is given by ionisation from the upper level. If, at this moment, the molecule were in the g-state, it will automatically be in the u-state half a laser period later and be ionised.

An interesting situation has now arisen. For small R, when the electron is in a molecular orbital, ionisation occurs from the ground level with no significant Stark shift. However, for large R, when the electron is localised, ionisation occurs from the upper level with a large Stark shift. Clearly, the

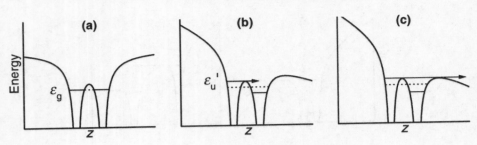

Fig. 2.24. (a) The central barrier localises the electron in either the left- or right-hand potential well. (b) The maximum upward shift is $\frac{1}{2}RF$, but never higher than the peak of the central barrier where the electron becomes delocalised. (c) At intermediate R, a further increase of the laser electric field is required for ionisation.

transition between these two situations will take place in the intermediate region of R, where the electron changes from being delocalised to being localised. The simple *field-ionisation, Coulomb-explosion* (FICE) model, which is consistent with the classical picture of over-the-barrier ionisation, was developed in order to describe this transition [74, 76, 77].

Classically, the electron is localised when the central barrier is higher than the unperturbed energy level ϵ_g, see figure 2.24 (a). When the laser is turned on, the localised electron is periodically shifted up and down with a maximum upward shift of $\frac{1}{2}RF$. However, the electron never shifts higher than the peak of the central barrier, since a delocalised electron does not experience a significant shift, see figure 2.24(b). At intermediate R, a further suppression of the outer barrier is necessary for the ionisation of the delocalised electron, see figure 2.24(c).

In order to calculate the threshold intensities of over-the-barrier ionisation as a function of R, we consider the following one-dimensional model of a molecular ion. The double-well potential, $U(x)$, in which the outer electron moves is modelled by [76]

$$U(x) = -\frac{\frac{1}{2}q}{|z + \frac{1}{2}R|} - \frac{\frac{1}{2}q}{|z - \frac{1}{2}R|} - Fz, \qquad (2.49)$$

where z is the axial coordinate of the electron. The energy level, ϵ_L, of the outer electron in this symmetric double well is approximated by

$$\epsilon_L = \frac{1}{2}(-\epsilon_1 - q_2/R) + \frac{1}{2}(-\epsilon_2 - q_1/R), \qquad (2.50)$$

where ϵ_1 and ϵ_2 are the known ionisation potentials of the atomic ions (e.g. $\epsilon_1 = \epsilon_2 = 0.5$ a.u., $q_1 = q_2 = 1$ and $q = 2$ for H_2^+). The maximum upward shift, which occurs only when the unperturbed level is below the

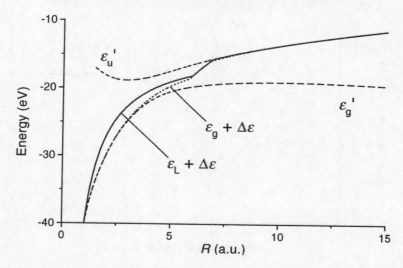

Fig. 2.25. The model energy level, $\varepsilon_L + \Delta\varepsilon$ (solid curve), follows the electronic ground state, ε'_g, for small R with reasonable accuracy and moves to the excited state, ε'_u, for larger R. For H_2^+ the model energy can be improved by using $\varepsilon_g + \Delta\varepsilon$ (the dotted curve).

central barrier, is equal to [77]

$$\Delta\epsilon_{\max} = \tfrac{1}{2}FR. \tag{2.51}$$

However, $\varepsilon_L + \Delta\varepsilon$ is limited by the height of the central barrier.

In figure 2.25, $\varepsilon_L + \Delta\varepsilon$ is compared with the Stark-shifted potentials E'_g and E'_u, which were discussed in section 2.4 on the dissociation of H_2^+ in the long-wavelength limit. Since these potentials include the Coulomb repulsion of the ion cores, the electronic energy of the two lowest states is given by $\varepsilon'_{g,u} = E'_{g,u} - q_1 q_2 / R$. We note that the model energy level $\varepsilon_L + \Delta\varepsilon$ (solid curve) indeed follows ε'_g for small R with reasonable accuracy and moves to ε'_u for larger R. These curves were produced with the parameters of H_2^+. For this molecule it is actually straightforward to improve the model energy by using ε_g instead of equation (2.50) (see the dotted curve).

Over-the-barrier appearance intensities are found from the minimum electric field that lowers the outer potential barrier below the (raised) electron-energy level of the parent ion. In figure 2.26 the appearance intensities for the (q_1, q_2) fragmentation channels of iodine are presented (full curves).

The appearance intensities of figure 2.26 clearly all have a deep minimum around $R = 10 a_0$. At this 'critical' internuclear separation, R_c, the threshold for ionisation is thus lower than it is for any other R. A molecular ion such as I_2, which is dissociating under the influence of an

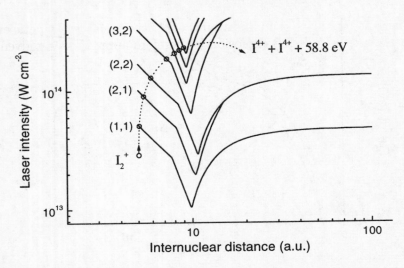

Fig. 2.26. The classical appearance intensities (solid curves) of the (q_1, q_2) fragmentation channels of I_2 have a minimum at the critical distance $R_c \approx 10a_0$. Classical trajectories (dotted curve) reproduce the experimental releases of energy in the Coulomb explosion with fair accuracy. The small circles indicate the instances of over-the-barrier ionisation. The laser pulse was Gaussian with a maximum of 2.5×10^{14} W cm^{-2} and a width of 150 fs FWHM (adapted from [60]).

intense laser field, will therefore most probably be further ionised when $R \approx R_c$. It is often said that the ionisation is *enhanced* at R_c.

Figure 2.26 further shows that the critical distances of the various channels are virtually identical. This is because the process of enhanced ionisation is closely related to the separation of the ion cores, which starts at roughly the same R for all outer-shell electrons [76]. By approximating the atomic ionisation potentials by $\varepsilon_q \approx q\varepsilon_I$, where ε_I is the first ionisation potential, it is in fact possible to get the analytical result [78]

$$R_c \approx \frac{4}{\varepsilon_I}, \tag{2.52}$$

which indeed yields $R_c \approx 10a_0$ for the I_2 molecule.

With the form of equation (2.50) it is implicitly assumed that the double potential wells are symmetric. However, when the ion cores are well separated this assumption is invalid for asymmetric channels such as (1,2). For large R this channel must be as difficult to create as the (2,2) channel, since both involve the appearance of an effectively free I^{2+} ion. The curves of the asymmetric channels are therefore allowed to approach smoothly the curves of the next higher symmetric channels between R_c and $R = 15a_0$.

Since the probability of ionisation is enhanced at R_c, the kinetic energy released by the Coulomb explosion is therefore approximately given by

$$E_c \approx \frac{q_1 q_2}{R_c}, \tag{2.53}$$

which explains why the kinetic energies of ions are virtually independent of pulse length and wavelength [79]. Still better agreement with the experimental results is achieved with classical trajectory calculations [77]. In this case, the evolutions of R and $I = I_0 \exp(-4\ln 2\, t^2/\Delta t^2)$ are followed through time. Over-the-barrier ionisation occurs when $I(t) \geq I_{app}(R)$ for the first time, where I_{app} stands for the appearance intensity. The evolution of $R(t)$ is given by the Coulomb repulsion. The trajectory for $I_0 = 2.5 \times 10^{14}$ W cm^{-2} and $\Delta t = 150$ fs results in the (4,4) channel with a kinetic energy release of 58.8 eV. This is to be compared with the experimental value of 60 eV, for which the margin of error is of the order of 5% [77]. Since both $R(t)$ and $I(t)$ are known, it is possible to portray the trajectory by plotting I against R, as illustrated in figure 2.26 by the dotted curve. Besides the fair agreement in absolute terms, the trajectories reproduce particularly well the small decrease in release of kinetic energy with increasing pulse length. When the laser pulses are much longer than the characteristic time for dissociation, post-dissociative ionisation (PDI) at $R \gg R_c$ may occur [80]. PDI is also easily interpreted in terms of the FICE model.

It appears that all diatomic molecules display features of enhanced ionisation when they dissociate in an intense laser field. In particular the following molecules have been investigated: I_2 [77, 81–83], Cl_2 [84], O_2 [79, 85], CO [79, 86, 87] and N_2 [9, 18, 77, 79, 88, 89]. However, the unambiguous interpretation of Coulomb-explosion spectra is severely impeded by the lack of understanding of the strong-field photodissociation of all molecules except H_2^+. Even dissociative recombination might contribute to the strong-field dissociation of complex molecules [90]. Naturally, the FICE model is also too simple to explain non-sequential ionisation of molecules [91].

In recent years, laser pulses have become so short that the inertial confinement restrains the fragmentation during the laser pulse, particularly in the case of I_2 [82]. This development has not only allowed the process of enhanced ionisation to be investigated with pump-probe techniques [81] but also led to new types of excitation. For example, ultra-short pulses produce the stable and metastable molecular ions, I_2^+, I_2^{2+} and I_2^{3+} [92]. Asymmetric-fragmentation channels, i.e. ones with $|q_1 - q_2| \geq 2$, are also related to fast rising pulse edges [77] (see also [93]). The observation of e.g. the (3,1) channel suggests that a Floquet state in the I_2^{4+} transient molecule is excited, in which two electrons slosh back and forth between two I^+ cores. The effect of the wavelength on charge-asymmetry is still under investigation [88, 94, 95].

Besides the heavier diatoms, experiments also indicate the existence of a critical distance in H_2^+, see [15, 21, 96, 97] and, with hindsight, [98]. We will see in the following sections that the quantum-mechanical treatment of strong-field ionisation of H_2^+ essentially underpins the ideas about enhanced ionisation that were presented in this section.

Quasi-static field ionisation of H_2^+

In this section we consider the field ionisation of H_2^+ with the molecular axis aligned along the field. In the long-wavelength limit, the laser interaction can be treated as a quasi-static electric field. The sinusoidal oscillations can be taken into account by time-averaging the rate of ionisation over one laser period [99]:

$$\bar{\Gamma} = \frac{2\omega}{\pi} \int_0^{\pi/(2\omega)} \Gamma\left(F\right) \mathrm{d}t. \tag{2.54}$$

It is possible to calculate the rate of ionisation for a time-independent problem with an eigenvalue method. Whereas only the strongly coupled $1\mathrm{s}\,\sigma_\mathrm{g}$ and $2\mathrm{p}\,\sigma_\mathrm{u}$ states were required for the strong-field dissociation, for the calculation of the rates of ionisation, special sets of basis functions are being used. Furthermore, the complex-basis method utilises a mixture of complex and real basis functions. Alternatively, with the complex-rotation method the basis remains real, but the Hamilton operator is made complex by rotating the coordinates into the complex plane. The basis functions are coupled by the molecular Hamiltonian, as well as the static field. The diagonalisation of the Hamilton matrix with the physical boundary conditions of outgoing waves yields complex eigenstates and eigenvalues. The eigenvalues of the system can be separated into real and imaginary parts according to

$$E = E_\mathrm{R} - \mathrm{i}\Gamma/2, \tag{2.55}$$

with E_R corresponding to the shifted energy and Γ being the rate of ionisation of the state. Two of the eigenfunctions correspond to the dressed states ϕ_g' and ϕ_u'. The time dependence of the eigenfunctions, given by $\mathrm{e}^{-\mathrm{i}Et}$, contains a real part and thus the norm of the eigenstates, $\langle\Psi^*|\Psi\rangle$, decays as $\mathrm{e}^{-\Gamma t}$.

The time-averaged rate of ionisation for the upper level, $\bar{\Gamma}_\mathrm{u}$, calculated with the complex-basis method with $F_0 = 0.04$ a.u., is illustrated in figure 2.27 (the $\bullet$ symbols). $\bar{\Gamma}_\mathrm{u}$ has two broad maxima around the critical distances $R_\mathrm{c} \approx 7a_0$ and $R_\mathrm{c} \approx 10a_0$. $\bar{\Gamma}_\mathrm{g}$ is considerably smaller and, furthermore, is a smooth function of R without any sign of critical distances.

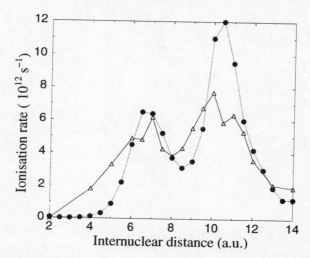

Fig. 2.27.   The rate of ionisation of H_2^+ as a function of R. A comparison between time-averaged results for DC field ionisation of ϕ_u with field strength $F_0 = 0.045$ a.u. [99] (•) and the time-dependent results of Zuo and Bandrauk [100] (△).

$\bar{\Gamma}_g$ is not plotted in figure 2.27; on the linear scale it would be virtually invisible.

The enhancement of Γ_u at *two* critical distances can be qualitatively understood in terms of the barrier model. For small R, ϕ'_u is still delocalised and ionisation proceeds via tunnelling through the outer barrier. For increasing R (keeping F fixed), the rate of ionisation initially increases due to the lowering of the outer barrier as well as the raising of the energy level. However, for $R > 5a_0$, the atomic orbitals cease to overlap and, whilst the ground state localises in the lower well, the electron cloud of ϕ'_u moves away from the outer barrier. Thus the first maximum in Γ_u is caused by a lowering of the effective barrier, counteracted by a retreat of the wavefunction to the opposite well.

Once the electron is completely localised in the upper well, the ionisation must proceed through (or across) both the central and the outer barrier. A further increase in R raises the inner barrier and lowers the outer. The last effect initially raises Γ_u once more, until the inner barrier rises to such an extent that it severely impedes the ionisation. The counteracting effects of the outer and inner barrier on the localised electron thus result in a second critical distance.

In the classical FICE model, which was discussed in the previous section, R_c originated from the hindrance of over-the-barrier ionisation by the inner barrier. It appears therefore that the classical critical distance corresponds to the second R_c of Γ_u. Whereas the classical FICE model considers only the top of the highest barrier with respect to the energy level,

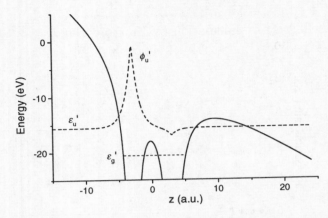

Fig. 2.28. For small R, ϕ'_u is still (partially) delocalised and ionisation proceeds via tunnelling through the outer barrier. A combination of electron localisation and barrier heights leads to two critical distances at which the rate of ionisation is enhanced.

the reality of quantum-mechanical ionisation turns out to be more complex. For a specific F, the maximum rate of ionisation usually occurs in a situation in which the outer barrier is much further suppressed than the central one. Consequently, the second R_c shifts towards smaller R with increasing laser intensity. Other subtleties that result from a quantum-mechanical treatment are the oscillations in Γ_u for large R, which are caused by above-barrier scattering and lead to large interference effects in the angular distribution of the ejected electrons [99].

We may ask why the FICE model overlooks the first R_c even though it includes the transition from a delocalised to a localised electron. The answer is that the FICE model assumes that the molecule is in the ground state as long the electron is delocalised. The first enhancement is thus suppressed. When H_2^+ dissociates in the long-wavelength limit, see section 2.4, the upper state will not be populated until the electron is localised. The FICE model, which is indeed a static model, is thus in essence consistent with the long-wavelength limit.

Charge-resonance-enhanced ionisation of H_2^+

We noted already in section 2.4 that the long-wavelength limit is not appropriate for a description of the strong-field dissociation of H_2^+ in the near-infrared region, i.e. in the wavelength ranges of the Ti : sapphire and Nd : YAG lasers. In this case, the nuclear dynamics is understood in terms of Floquet states. The probabilities of ionisation from these states can be found by wavepacket calculations [100, 101]. The rate of ionisation is found from the exponential decay of the norm of the wavefunction integrated over the whole volume.

The result of such a calculation by Zuo and Bandrauk [100] for $I = 10^{14}$ W cm^{-2} and $\lambda = 1064$ nm is plotted in figure 2.27 (the $\triangle$ symbols) and compared with the time-averaged, quasi-static field-ionisation rates of ϕ'_u [99] (the $\bullet$ symbols). Besides having two critical distances, these rates are also in quantitative terms remarkably similar.

For large R, the central barrier impedes the transfer of the electron from one well to the other and thus the g- and u-states switch continuously (Rabi flopping). For symmetry reasons, the two states in the dynamic model are thus both 50% populated and this appears to be reflected in the rates of ionisation around the second critical distance.

In the discussion on vibrational trapping, we noted that if the initial R is beyond the one-photon crossing of the dressed states (i.e. for R larger than $\simeq 5a_0$ at $\lambda = 1064$ nm) the g-state evolves into the upper branch of the Floquet states. In the upper Floquet state, the electron moves in such a way that it is predominantly on the raised side of the potential well, just like the u-state from the quasi-static model. So, in the dynamic model, the first R_c is due to the charge-resonance excitation of the ground state. At high intensities, such as 10^{14} W cm^{-2}, the rate of *charge-resonance-enhanced ionisation* (CREI) is approximately given by that of the u-state of the quasi-static model. At lower intensities, the relative importance of CREI diminishes.

If the wavelength is made shorter, the one-photon crossing shifts to smaller R and, consequently, so does the critical distance due to CREI. The second R_c, however, also shifts towards smaller R. We recall that the second R_c was related to tunnelling through the central barrier directly into the continuum. This process is significantly suppressed at higher frequencies, since the laser electric field does not remain sufficiently long at its maxima for a tunnelling current to build up. Consequently, the central barrier becomes more effective at localising the electron, resulting in a smaller second R_c. At 600 nm and 10^{14} W cm^{-2}, for example, the two critical distances have merged into one broad range of enhanced ionisation, peaking at $R \approx 8a_0$ [102].

On the lower branch of the one-photon crossing, the electron cloud moves in phase with the electric field. At high intensities (i.e. when the crossing becomes adiabatic and therefore active) and at the crests and troughs of the field, the electron is predominantly on the lowered side of the potential well. In this case, the CREI mechanism will presumably be suppressed. Calculations that also treat the nuclei quantum-mechanically give more insight into the effect of the dissociation dynamics on the process of enhanced ionisation [103].

In figure 2.29 two of the spectra that were already presented in figure 2.16 are plotted on an energy scale. We noted in section 2.4 that the dom-

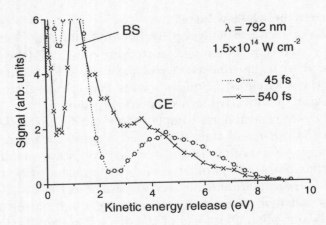

Fig. 2.29. Kinetic-energy-release spectra of H_2 subjected to laser pulses of 45 and 540 fs FWHM, showing the Coulomb-explosion (CE) and bond-softening (BS) channels.

inant (0,1) channel is due to two-photon bond-softening (BS), whereas the other is interpreted as below-threshold dissociation (BTD). We expect the CREI mechanism to work in particular on the upper branch of the one-photon crossing (and to some extent also at higher-order crossings). The BS wavepacket follows the lower branch at the three-photon crossing and indeed the upper branch at the one-photon crossing. Assuming that the kinetic energy released is given by

$$E_m = E_{BS} + \frac{e^2}{R}, \qquad (2.56)$$

where $E_{BS} \approx 1$ eV, the energy of the (1,1) peak of the $\Delta t = 45$ fs spectrum suggests that $R_c \approx 8a_0$ at 794 nm. However, it is clear that the peak shifts to lower energies when the laser pulses are lengthened. At $\Delta t = 540$ fs we find that $R_c \approx 10a_0$. It appears that the process of enhanced ionisation is biased towards short R for ultra-short pulses due to inertial confinement. This effect is even stronger with the heavier isotope D_2 [97]. The shoulder, which has developed on the high-energy side of the BS peak at 540 fs, is probably due to Coulomb explosion well beyond the critical region (post-dissociative ionisation).

The calculations of Charron and coworkers [36] that were discussed in section 2.4 (see also figure 2.9) predict that the BTD peak is related to vibrational trapping around $R \approx 4a_0$, see figure 2.17. Since the vibrational excitation in the light-induced well is minimal, ionisation of the trapped state should yield a kinetic-energy release of $\simeq 6.8$ eV. The high-energy wing of the Coulomb-explosion peaks in figure 2.29 could thus be due to ionisation of the trapped state. Presumably the two channels are not resolved because of the large widths of the individual peaks.

2.6 Conclusions

We have reviewed some elementary experimental and theoretical aspects of diatomic molecules in intense laser fields. Photodissociation and Coulomb explosion are usually studied by ion time-of-flight techniques. Important experimental considerations in the design of the interaction chamber should be energy resolution, collection efficiency, target cooling and space charge. The distortion of the relative weight of phenomena by focal-volume effects is easily underestimated. A weak signal is best optimised not by increasing the laser intensity, but by increasing the laser power and changing to a longer focal length. A further improvement is brought about by screening off most of the ions from the outer shells. This is only possible with a large f-ratio. One of the experiments that should greatly benefit from such a screening is the study of the intensity windows in which certain fragmentation channels occur. This type of experiment was discussed in section 2.3, together with the Keldysh parameter, which assists in the classification of ionisation into the multi-photon, the tunnelling and the over-the-barrier regimes.

Bond-softening and above-threshold dissociation are two typical processes associated with the strong-field photodissociation of H_2^+. Numerical integration of the time-dependent Schrödinger equation can, in principle, give excellent agreement with experimental results. For physical insight, a quasi-static model, which describes strong-field dissociation in the long-wavelength limit, is very instructive. At shorter wavelengths, the Floquet method and the photon-dressed states are highly effective for the understanding of vibrational trapping and other peculiar phenomena.

The strong-field ionisation of diatomic molecules was discussed in terms of both quasi-static and dynamic quantum-mechanical models and even a classical model. The various approaches all predict an enhancement of the probability of ionisation at critical distances. Further analysis suggests that these are related to localisation of the electron on the high-energy side of the molecule. The associated excitation results both from a quasi-static localisation by a high inner barrier and from charge-resonance excitation to a Floquet state in which the electron cloud oscillates in antiphase with the external field. The enhanced ionisation is strongly affected by the dynamics of the dissociation, as illustrated, for example, by the observation of Hering and Cornaggia [104], who found that the rate of production of multiply charged atoms from molecules with equivalent atoms, such as N_2, is one order of magnitude greater than that from molecules built with a single C, N or O atom, such as NH_3.

We have concentrated on the core aspects of diatomic molecules in intense laser fields and, for example, completely omitted the subject of angular distributions. The Coulomb explosion of diatomic molecules invariably leads to ion angular distributions that are strongly peaked along the direction of the electric vector of the linearly polarised laser field. To a large extent this is due to a geometrical effect. For a specific laser electric field, the enhancement of ionisation is strongest for molecules aligned along the axis of polarisation. Indeed, when the polarisation is perpendicular to the molecular axis, the process of ionisation is very atom-like [74, 81]. For bond-softening and other photodissociation phenomena in H_2^+, similar arguments hold. As in many theoretical studies, we have therefore implicitly assumed throughout this chapter that the molecular axis and the laser electric field are always parallel.

Recent studies [105] indicate that the peaked angular distributions of Coulomb explosion of I_2, subjected to ultra-short Ti : sapphire laser pulses, are indeed due solely to a variation of the probabilities of ionisation with angle. With lighter molecules, however, further dynamic alignment adds to these geometrical effects. The laser electric field and the dipole that it induces combine to give a torque that tends to align the molecular axis. In the case of a rigid rotor (fixed R), this leads to pendular states [106]. Many interesting contributions to the subject of laser-induced alignment of diatoms have been made, both experimentally [15, 82, 87, 107] and theoretically [108].

In the infrared-wavelength regime, it has often been assumed that the process of ionisation of the neutral H_2 molecule does not interfere with the subsequent dynamics of H_2^+. The comparison of experimental results on H_2 with theoretical work on H_2^+ seemed, therefore, warranted. On the other hand, the participation of excited electronic states in the multiphoton ionisation of H_2 has long been accepted for relatively long laser pulses in the ultraviolet-wavelength regime [8, 109]. Nevertheless, in the case of ionisation by ultra-short laser pulses at 800 nm, the distribution of the population amongst the vibrational levels of H_2^+ was, quite naturally, generally considered to be given by the Franck–Condon overlap with the ground level of neutral H_2 [110] (the Frank–Condon factors for the ionisation of H_2 and D_2 can be found in [111]; see also [112]). Recent work indicates, however, that a lower range of vibrational levels is populated [22, 50]. It appears that the Franck–Condon principle does not apply due to the profound variation of the tunnelling amplitude with internuclear separation. Furthermore, some form of bond-softening seems to take place also in the neutral molecule [113] and the ionisation of H_2 thus occurs at a relaxed internuclear separation. The $v = 0$ level of H_2^+ is expected to be highly populated but its probability of tunnelling through the bond-softened barrier is small. This prevents enhanced ionisation at larger R.

Intensities above 10^{15} W cm^{-2} are required before these molecules show up as a high-energy shoulder in the Coulomb-explosion channel [114].

Clearly, a simple modelling of appearance intensities, such as the one that we presented in section 2.3, cannot take into account the different rates of ionisation and dissociation of the various vibrational levels. Ideally, H_2^+ should be produced in pure vibrational states before its interaction with a strong laser field is investigated. In principle this can be achieved with resonance-enhanced multi-photon ionisation of H_2. Electron spectra show that it is possible to produce practically pure, low vibrational levels [8, 115, 116]. Another interesting line of investigation was followed by Rottke and coworkers [49], who frequency tripled the radiation from a dye laser in order to prepare excited H_2 molecules in a pure rovibrational level of the B-state. These methods with tunable, short wavelength pulses will, it is to be hoped, soon be transferred to the femtosecond regime. This should lead to new opportunities to study and influence the dynamics of multi-photon excitation both of H_2 and H_2^+.

Acknowledgements

We acknowledge the UK Engineering and Physical Sciences research council for their financial support. We are also grateful to Professors Codling, Connerade and Walther for their having given us so much assistance and encouragement.

References

[1] K. Codling and L. J. Frasinski, *J. Phys. B* , **26** 783 (1993).

[2] A. L'Huillier, L. Lompré, G. Mainfray and C. Manus, *Atoms in Intense Laser Fields*, ed. M. Gavrila, Academic Press, Boston, 1992.

[3] K. Burnett, V. C. Reed and P. L. Knight, *J. Phys. B* **26** 561 (1993).

[4] T. Ditmire, T. Donnelly, A. M. Rubenchik, R. W. Falcone and M. D. Perry, *Phys. Rev. A* **53** 3379 (1996).

[5] J. W. G. Tisch, N. Hay, E. Springate, E. T. Gumbrell, M. H. R. Hutchinson and J. P. Marangos, *Phys. Rev. A* **66** 3076 (1999).

[6] R. S. Mulliken, *J. Chem. Phys.* **7** 20 (1939).

[7] M. Brewczyk, K. Rzążewski and C. W. Clark, *Phys. Rev. Lett.* **78** 191 (1997).

[8] C. Cornaggia, D. Normand, J. Morellec, G. Mainfray and C. Manus, *Phys. Rev. A* **34** 207 (1986).

[9] L. J. Frasinski, K. Codling, P. A. Hatherly, J. Barr, I. N. Ross and W. T. Toner, *Phys. Rev. Lett.* **58** 2424 (1987).

[10] P. H. Bucksbaum, A. Zavriyev, H. G. Muller and D. W. Schumacher, *Phys. Rev. Lett.* **64** 1883 (1990).

[11] S. L. Chin, Y. Liang, J. E. Decker, F. A. Ilkov and M. V. Ammosov, *J. Phys. B* **25** L249 (1992).

[12] G. Ravindra Kumar, C. P. Safvan, F. A. Rajgara and D. Mathur, *J. Phys. B* **27** 2981 (1994).

[13] P. Kruit and F. H. Read *J. Phys. E* **16** 313 (1983).

[14] W. C. Wiley and I. H. McLaren, *Rev. Sci. Instrum.* **26** 1150 (1955).

[15] M. R. Thompson, M. K. Thomas, P. F. Taday, J. H. Posthumus, A. J. Langley, L. J. Frasinski and K. Codling, *J. Phys. B* **30** 5755 (1997).

[16] L. J. Frasinski, K. Codling and P. A. Hatherly, *Science* **246** 1029 (1989).

[17] J. Zhu and W. T. Hill III, *J. Opt. Soc. Am. B* **14** 2212 (1997).

[18] A. Hishikawa, A. Iwamae, K. Hoshina, M. Kono and K. Yamanouchi, *Chem. Phys.* **231** 315 (1998).

[19] C. Cornaggia, J. Lavancier, D. Normand, J. Morellec and H. X. Liu, *Phys. Rev. A* **42** 5464 (1990).

[20] J. E. Decker, G. Xu and S. L. Chin, *J. Phys. B* **24** L281 (1991).

[21] J. Ludwig, H. Rottke and W. Sandner, *Phys. Rev. A* **56** 2168 (1997).

[22] H. Rottke, C. Trump and W. Sandner, *Laser Phys.* **9** 171 (1999).

[23] A. E. Siegman, *Lasers*, Oxford University Press, Oxford, 1986.

[24] P. Hansch, M. A. Walker and L. D. Van Woerkom, *Phys. Rev. A* **54** R2559 (1996); P. Hansch and L. D. Van Woerkom *Opt. Lett.* **21** 1286 (1996).

[25] S. Dobosz, M. Lewenstein, M. Lezius, D. Normand and M. Schmidt, *J. Phys. B* **30** L757 (1997).

[26] I. D. Williams, P. McKenna, B. Srigengan, I. M. G. Johnston, W. A. Bryan, J. H. Sanderson, A. El-Zein, T. R. J. Goodworth, W. R. Newell, P. F. Taday and A. J. Langley, *J. Phys. B* **33** 2743 (2000).

[27] K. Sändig, H. Figger and T. W. Hänsch, *Phys. Rev. Lett.* **85** 4876 (2000).

[28] C. Wunderlich, E. Kobler, H. Figger and T. W. Hänsch, *Phys. Rev. Lett.* **78** 2333 (1997).

[29] L. V. Keldysh, *Sov. Phys. JETP* **20** 1307 (1965).

[30] A. M. Peremolov, V. S. Popov and M. V. Terent'ev, *Sov. Phys. JETP* **23** 924 (1966).

[31] M. V. Ammosov, N. B. Delone and V. P. Krainov, *Sov. Phys. JETP* **64** 1191 (1986).

[32] V. P. Krainov, *J. Nonlinear Opt. Phys. Mater.* **4** 775 (1995).

[33] D. Bauer and P. Mulser, *Phys. Rev. A* **59** 569 (1999).

[34] R. R. Freeman and P. H. Bucksbaum, *J. Phys. B* **24** 325 (topical review) (1991); L. F. DiMauro and P. Agostini, in *Advances in Atomic, Molecular, and Optical Physics* Vol. 35, ed. B. Bederson and H. Walther, Academic Press, San Diego, 1995.

[35] H. G. Muller, in *Coherence phenomena in Atoms and Molecules in Laser Fields*, ed. A. D. Bandrauk and S. C. Wallace, Plenum, New York, 1992 (*NATO ASI Series* B **287** 89).

[36] A. Giusti-Suzor, F. H. Mies, L. F. DiMauro, E. Charron and B. Yang, *J. Phys. B* **28** 309–339 (topical review) (1995).

[37] A. D. Bandrauk and M. L. Sink, *J. Chem. Phys.* **74** 1110 (1981).

[38] J. R. Hiskes, *Phys. Rev.* **122** 1207 (1961).

[39] P. Dietrich and P. B. Corkum, *J. Chem. Phys.* **97** 3187 (1992).

[40] T. D. G. Walsh, L. Strach and S. L. Chin, *J. Phys. B* **31** 4853 (1998).

[41] A. Giusti-Suzor, X. He, O. Atabek and F. H. Mies, *Phys. Rev. Lett.* **64** 515 (1990).

[42] D. R. Bates, *J. Chem. Phys.* **19** 1122 (1951).

[43] H. B. van Linden van den Heuvell and H. G. Muller, in *Multiphoton Processes, Cambridge Studies in Modern Optics* Vol. 8, ed. S. J. Smith and P. L. Knight, Cambridge University Press, New York, 1988.

[44] P. Dietrich, M. Yu. Ivanov, F. A. Ilkov and P. B. Corkum, *Phys. Rev. Lett.* **77** 4150 (1996).

[45] J. H. Shirley, *Phys. Rev.* **138** B979 (1965).

[46] S.-I Chu, *J. Chem. Phys.* **75** 2215 (1981); T. F. George, *J. Phys. Chem.* **86** 10 (1982); X. He, O. Atabek and A. Giusti-Suzor, *Phys. Rev. A* **38** 5586 (1988).

[47] R. J. Glauber, *Phys. Rev.* **131** 2766 (1963).

[48] G. Jolicard and O. Atabek, *Phys. Rev. A* **46** 5845 (1992).

[49] H. Rottke, J. Ludwig and W. Sandner, *J. Phys. B* **30** 2835 (1997).

[50] L. J. Frasinski, J. H. Posthumus, J. Plumridge, K. Codling, P. F. Taday and A. J. Langley, *Phys. Rev. Lett.* **83** 3625 (1999).

[51] A. Giusti-Suzor and F. H. Mies, *Phys. Rev. Lett.* **68** 3869 (1992).

[52] G. Yao and S.-I Chu, *Chem. Phys. Lett.* **197** 413 (1992).

[53] R. Numico, A. Keller and O. Atabek, *Phys. Rev. A* **56** 772 (1997).

[54] J. H. Posthumus, J. Plumridge, L. J. Frasinski, K. Codling, E. Divall, A. J. Langley and P. F. Taday, *J. Phys. B* **33** L563 (2000).

[55] F. H. M. Faisal, *J. Phys. B* **6** L89 (1973).

[56] H. R. Reiss, *Phys. Rev. A* **22** 1786 (1980).

[57] F. Trombetta, S. Basile and G. Ferrante, *J. Mod. Opt.* **36** 891 (1989).

[58] M. Dörr, R. M. Potvliege, D. Proulx and R. Shakeshaft, *Phys. Rev. A* **42** 4138 (1990).

[59] B. H. Bransden and C. J. Joachain, *Physics of Atoms and Molecules*, Longman, London, 1983.

[60] J. F. McCann and J. H. Posthumus, *Phil. Trans. R. Soc.* **357** 1309 (1999).

[61] D. Dundas, J. F. McCann, J. S. Parker and K. T. Taylor, *J. Phys. B* **33** 3261 (2000).

[62] S. Vivirito, K. T. Taylor and J. S. Parker, *J. Phys. B* **31** L711 (1998).

[63] S. Vivirito, K. T. Taylor and J. S. Parker, *J. Phys. B* **32** 3015 (1999).

[64] M. Plummer, J. F. McCann and L. B. Madsen, *Comput. Phys. Commun.* **114** 94 (1998).

[65] L. B. Madsen, M. Plummer and J. F. McCann, *Phys. Rev. A* **58** 456 (1998).

[66] M.-G. Baik, M. Pont and R. Shakeshaft, *Phys. Rev. A* **54** 1570 (1996).

[67] L. B. Madsen and M. Plummer, *J. Phys. B* **31** 87 (1998).

[68] S. Chelkowski, S. Zuo and A. Bandrauk, *Phys. Rev. A* **46** R5342 (1992).

[69] J. Colgan, D. H. Glass, K. Higgins and P. G. Burke, *Comput. Phys. Commun.* **114** 27 (1998).

[70] P. G. Burke, J. Colgan, D. H. Glass and K. Higgins, *J. Phys. B* **33** 143 (2000).

[71] A. Apelalategui and P. Lambropoulos, *J. Phys. B* **33** 2791 (2000).

[72] C. P. Safvan and D. Mathur, *J. Phys. B* **27** 4073 (1994).

[73] K. Codling, L. J. Frasinski, P. A. Hatherly and J. R. M. Barr, *J. Phys. B* **20** L525 (1987).

[74] K. Codling, L. J. Frasinski and P. A. Hatherly, *J. Phys. B* **22** L321 (1989).

[75] S. Augst, D. Strickland, D. D. Meyerhofer, S. L. Chin and J. H. Eberly, *Phys. Rev. Lett.* **63** 2212 (1989); S. Augst, D. D. Meyerhofer, D. Strickland and S. L. Chin, *J. Opt. Soc. Am. B* **8** 858 (1991).

[76] J. H. Posthumus, L. J. Frasinski, A. J. Giles and K. Codling, *J. Phys. B* **28** L349 (1995).

[77] J. H. Posthumus, A. J. Giles, M. R. Thompson, L. J. Frasinski, K. Codling, A. J. Langley and W. Shaikh, *J. Phys. B* **29** L525 (1996); J. H. Posthumus, A. J. Giles, M. R. Thompson and K. Codling, *J. Phys. B* **29** 5811 (1996).

[78] S. Chelkowski and A. D. Bandrauk, *J. Phys. B* **28** L723 (1995); H. T. Yu, T. Zuo and A. D. Bandrauk, *J. Phys. B* **31** 1533 (1998).

[79] C. Cornaggia, J. Lavancier, D. Normand, J. Morellec, P. Agostini, J. P. Chambaret and A. Antonetti, *Phys. Rev. A* **44** 4499 (1991).

[80] M. Schmidt, P. D'Oliveire, P. Meynadier, D. Normand and C. Cornaggia, *J. Nonlinear Opt. Phys. Mater.* **4** 817 (1995).

[81] E. Constant, H. Stapelfeldt and P. B. Corkum, *Phys. Rev. Lett.* **76** 4140 (1996).

[82] D. T. Strickland, Y. Beaudoin, P. Dietrich and P. B. Corkum, *Phys. Rev. Lett.* **68** 2755 (1992).

[83] P. Dietrich, D. T. Strickland and P. B. Corkum, *J. Phys. B* **26** 2323 (1993); P. A. Hatherly, M. Stankiewicz, K. Codling, L. J. Frasinski and G. M. Cross, *J. Phys. B* **27** 2993 (1994); D. Normand and M. Schmidt, *Phys. Rev. A* **53** R1958 (1996).

[84] M. Schmidt, D. Normand and C. Cornaggia, *Phys. Rev. A* **50** 5037 (1994).

[85] D. Normand and C. Cornaggia, J. Lavancier, J. Morellec and H. X. Liu, *Phys. Rev. A* **44** 475 (1991).

[86] J. Lavancier, D. Normand, C. Cornaggia, J. Morellec and H. X. Liu, *Phys. Rev. A* **43** 1461 (1991).

[87] P. A. Hatherly, L. J. Frasinski, K. Codling, A. J. Langley and W. Shaikh, *J. Phys. B* **23** L291 (1990).

[88] K. Codling, C. Cornaggia, L. J. Frasinski, P. A. Hatherly, J. Morellec and D. Normand, *J. Phys. B* **24** L593 (1991).

[89] C. Cornaggia, D. Normand and J. Morellec, *J. Phys. B* **25** L415 (1992).

[90] A. Talebpour, C.-Y. Chien and S. L. Chin, *J. Phys. B* **29** L677 (1996).

[91] A. Talebpour, S. Larochelle and S. L. Chin, *J. Phys. B* **30** L245 (1997); C. Cornaggia and Ph. Hering, *J. Phys. B* **31** L503 (1998); C. Guo, M. Li, J. P. Nibarger and G. N. Gibson, *Phys. Rev. A* **58** R4271 (1998).

[92] H. Sakai, H. Stapelfeldt, E. Constant, M. Y. Ivanov, D. R. Matusek, J. S. Wright and P. B. Corkum, *Phys. Rev. Lett.* **81** 2217 (1998).

[93] G. N. Gibson, M. Li, C. Guo and J. P. Nibarger, *Phys. Rev. A* **58** 4723 (1998).

[94] K. Boyer, T. S. Luk, J. C. Solem and C. K. Rhodes, *Phys. Rev. A* **39** 1186 (1989).

[95] W. T. Hill III, J. Zhu, D. L. Hatten, Y. Cui, J. Goldhar and S. Yang, *Phys. Rev. Lett.* **69** 2646 (1992).

[96] F. A. Ilkov, S. Turgeon, T. D. G. Walsh and S. L. Chin, *Chem. Phys. Lett.* **247** 1 (1995); Y. L. Shao, D. J. Fraser, M. H. R. Hutchinson, J. Larsson, J. P. Marangos and J. W. G. Tisch, *J. Phys. B* **29** 5421 (1996); G. N. Gibson, M. Li, C. Guo and J. Neira, *Phys. Rev. Lett.* **79** 2022 (1997); C. Trump, H. Rottke and W. Sandner, *Phys. Rev. A* **59** 2858 (1999); J. H. Posthumus, J. Plumridge, P. F. Taday, J. H. Sanderson, A. J. Langley, K. Codling and W. A. Bryan, *J. Phys. B* **32** L93 (1999).

[97] T. D. G. Walsh, F. A. Ilkov and S. L. Chin, *J. Phys. B* **30** 2167 (1997).

[98] K. Codling, L. J. Frasinski and P. A. Hatherly, *J. Phys. B* **21** L433 (1988); T. S. Luk and C. K. Rhodes, *Phys. Rev. A* **38** 6180 (1988); A. Zavriyev,

P. H. Bucksbaum, J. Squier and D. W. Schumacher, *Phys. Rev. Lett.* **70** 1077 (1993).

[99] M. Plummer and J. F. McCann, *J. Phys. B* **29** 4625 (1996).

[100] T. Zuo and A. D. Bandrauk, *Phys. Rev. A* **52** R2511 (1995).

[101] T. Seideman, M. Yu. Ivanov and P. B. Corkum, *Phys. Rev. Lett.* **75** 2819 (1995); C. Leforestier and R. E. Wyatt, *J. Chem. Phys.* **78** 2334 (1983); D. Neuhasuer and M. Baer, *J. Chem. Phys.* **90** 4351 (1989).

[102] S. Chelkowski, A. Conjusteau, T. Zuo and A. D. Bandrauk, *Phys. Rev. A* **54** 3235 (1996).

[103] S. Chelkowski, T. Zuo, O. Atabek and A. D. Bandrauk, *Phys. Rev. A* **52** 2977 (1995); K. C. Kulander, F. H. Mies and H. J. Schafer, *Phys. Rev. A* **53** 2562 (1996); S. Chelkowski, C. Foisy and A. D. Bandrauk, *Phys. Rev. A* **57** 1176 (1998).

[104] Ph. Hering and C. Cornaggia, *Phys. Rev. A* **57** 4572 (1998).

[105] J. H. Posthumus, J. Plumridge, M. K. Thomas, K. Codling, L. J. Frasinski, A. J. Langley and P. F. Taday, *J. Phys. B* **31** L553 (1998); Ch. Ellert, H. Stapelfeldt, E. Constant, H. Sakai, J. Wright, D. M. Rayner and P. B. Corkum, *Phil. Trans. R. Soc. A* **356** 329 (1998); C. Ellert and P. B. Corkum, *Phys. Rev. A* **59** R3170 (1999).

[106] B. Friedrich and D. Herschbach, *Phys. Rev. Lett.* **74** 4623 (1995); W. Kim and P. M. Felker, *J. Chem. Phys.* **104** 1147 (1996); B. Friedrich and D. Herschbach, *Chem. Phys. Lett.* **262** 41 (1996).

[107] A. Zavriyev, P. H. Bucksbaum, H. G. Muller and D. W. Schumacher, *Phys. Rev. A* **42** 5500 (1990); D. Normand, L. A. Lompré and C. Cornaggia, *J. Phys. B* **25** L497 (1992); P. Dietrich, D. T. Strickland, M. Laberge and P. B. Corkum, *Phys. Rev. A* **47** 2305 (1993); J. H. Posthumus, J. Plumridge, L. J. Frasinski, K. Codling, A. J. Langley and P. F. Taday, *J. Phys. B* **31** L985 (1998); J. J. Larsen, I. Wendt-Larsen and H. Stapelfeldt, *Phys. Rev. Lett.* **83** 1123 (1999); S. Banerjee, G. Ravindra Kumar and D. Mathur, *Phys. Rev. A* **60** R3369 (1999). M. Schmidt, S. Dobosz, P. Meynadier, P. D'Oliveira, D. Normand, E. Charron and A. Suzor-Weiner, *Phys. Rev. A* **60** 4706 (1999).

[108] J. F. McCann and A. D. Bandrauk, *J. Chem. Phys.* **96** 903 (1993); E. E. Aubanel, J. M. Gauthier and A. D. Bandrauk, *Phys. Rev. A* **48** 2145 (1993); E. E. Aubanel, A. Conjusteau and A. D. Bandrauk *Phys. Rev. A* **48** R4011 (1993); E. Charron, A. Giusti-Suzor and F. H. Mies, *Phys. Rev. A* **49** R641 (1994); R. Numico, A. Keller and O. Atabek, *Phys. Rev. A* **52** 1298 (1995); M. Plummer and J. F. McCann, *J. Phys. B* **30** L401 (1997); M. E. Sukharev and V. P. Krainov, *Sov. Phys. JETP* **86** 318 (1998).

[109] H. Helm, M. J. Dyer and H. Bissantz, *Phys. Rev. Lett.* **67** 1234 (1991).

[110] T. D. G. Walsh, F. A. Ilkov, S. L. Chin, F. Châteauneuf, T. T. Nguyen-Dang, S. Chelkowski, A. D. Bandrauk and O. Atabek, *Phys. Rev. A* **58** 3922 (1998).

[111] G. H. Dunn, *J. Chem. Phys.* **44** 2592 (1966).

[112] M. E. Sukharev and V. P. Krainov, *Sov. Phys. JETP* **83** 457 (1996).

[113] A. Saenz, *Phys. Rev. A* **61** 051402(R) (2000).

[114] C. Trump, H. Rottke and W. Sandner, *Phys. Rev. A* **60** 3924 (1999).

[115] E. Y. Xu, T. Tsuboi, R. Kachru and H. Helm, *Phys. Rev. A* **36** 5645 (1987).

[116] S. W. Allendorf and A. Szöke, *Phys. Rev. A* **44** 518 (1991).

3

Small polyatomic molecules in intense laser fields

Christian Cornaggia

CEA Saclay
Service des Photons, Atomes et Molécules
Division des Sciences de la Matière
F-91 191 Gif-sur-Yvette, France

3.1 Introduction

The multiple ionisation of small molecules is being studied with various excitation sources, e.g. ionisation by ion or electron impact, synchrotron radiation, beam-foil electron stripping and intense laser pulses. Since multiply charged molecular ions are unstable, the process of ionisation itself is investigated by analysing the multiply charged atomic fragments. The multiple fragmentation channels are determined using various experimental techniques adapted to the excitation sources. These methods present some similarities to other fields in physics, such as particle and nuclear physics, in which the fragmentation of particles and nuclei plays an important role. This chapter starts with some general considerations about the laser excitation of small molecules and then continues to discuss the physics and the experimental techniques that are associated with laser-induced multiple ionisation. The physical quantities are expressed in atomic units (a.u.), MKSA units and practical units such as W cm^{-2} for the laser intensity. Most of the atomic- and molecular-spectroscopy data are taken from standard textbooks, for instance the books of Herzberg for neutral and singly charged molecules [1, 2]. The transient molecular ions, which remain undetected but appear as the precursors of the observed multiple fragmentation channels, are noted in square brackets, for example [N_2O^{9+}].

Timescales

The laser field presents basically two timescales: the optical period $T = \lambda/c$, where λ and c are, respectively, the laser wavelength and the speed of light, and the duration of the laser pulse. The optical period T is of the order of one femtosecond for wavelengths in the near-ultraviolet, the visible and the near-infrared ranges ($T = 2$ fs at $\lambda = 600$ nm). The pulse duration lies in the 30 fs–100 ps range, depending on the gain medium of the laser.

These timescales have to be compared with the molecular electronic, vibrational and rotational timescales. For the excitation of neutral molecules from their electronic ground states, the electronic response time is of the order of 1 a.u. (1 a.u. $= 2.419 \times 10^{-2}$ fs) and is much shorter than the laser timescales. Therefore, laser experiments do not provide any direct insight into the temporal dynamics of the electron ejection. In the absence of a laser field, the nuclear motions are characterised by the rotational and vibrational frequencies. The associated timescales depend both on the molecular electronic state and on the masses of the nuclei. For H_2, the rotational and vibrational times are, respectively, 1 ps and 7.6 fs. For a linear triatomic molecule such as CO_2, the rotation time is 85 ps, whereas the vibrational modes exhibit different periods of 24 fs and 14 fs for the symmetric and antisymmetric stretching modes and 50 fs for the bending mode. For the shortest available laser pulses in the 30–100-fs pulse-duration range, the zero-field rotation time is thus always much longer than the duration of the laser pulse, whilst the vibrational times are of the same order of magnitude. Therefore, the molecule does not have time to rotate by itself. However, as we will see below, the laser-induced polarisability and the associated multiple ionisation effects significantly modify the rotational behaviour in comparison with the zero-field characteristics of the molecule. Since the vibrational times are of the order of the duration of the laser pulse, the molecule can vibrate during the multiple-ionisation process. This point will be discussed for linear species.

The fragmentation is in general very fast, since it is governed by the Coulomb repulsion of the charged fragments. Let us consider the simplest case, in which two atomic multi-charged atoms $A_1^{Z_1+}$ and $A_2^{Z_2+}$ begin to repel at some distance R_0. Assuming a pure Coulomb repulsion, $Z_1 Z_2/R$, where R is the internuclear distance, the time $T_{R_0}(R)$ taken to reach the distance R from R_0 is given by $T_{R_0}(R) = \int_{R_0}^{R} dR'/v(R')$, where $v(R')$ is the velocity at $R = R'$. This yields

$$T_{R_0}(R) = \frac{R_0}{v(R' = \infty)} \left[\sqrt{\frac{R}{R_0}} \sqrt{\frac{R}{R_0} - 1} + \ln\left(\sqrt{\frac{R}{R_0}} + \sqrt{\frac{R}{R_0} - 1} \right) \right], \quad (3.1)$$

where ln is the natural logarithm and $v(R' = \infty)$ is the velocity of the repelling system at $R' = \infty$, i.e. at the end of the dissociation. The characteristic time for repulsion in equation (3.1) is $R_0/v(R' = \infty)$ and can be expressed in femtosecond units as

$$\frac{R_0}{v(R' = \infty)} = 0.730 \left(\frac{M}{Z_1 Z_2} \right)^{1/2} \left(\frac{R_0}{a_0} \right)^{3/2}, \quad (3.2)$$

where M is the reduced mass (a.m.u.) of the repelling system and a_0 is the Bohr radius. This time is 0.855 fs for the $H^+ + H^+$ channel starting at the equilibrium internuclear distance of H_2, $R_0 = 0.741$ Å, and 2.89 fs for the $N^{2+} + N^{2+}$ channel starting at the equilibrium internuclear distance of N_2, $R_0 = 1.098$ Å. In equation (3.1), this time is multiplied by a function that depends only on the ratio R/R_0. This function behaves like R/R_0 for large R and, in consequence, the time for dissociation is of the order of one or several tens of femtoseconds.

In conclusion, with the present state of the art in femtosecond-laser technology, the process of fragmentation has in principle enough time to occur within the duration of the laser pulse. This specific feature of the laser interaction is different from beam–foil and fast-ion-beam experiments, in which the duration of interaction of the molecule either in the foil or with the impinging ion is only a fraction of a femtosecond [3–5]. In this case, following equation (3.1), the nuclear coordinates remain practically unchanged and equal to the equilibrium coordinates during the process of multiple ionisation.

The laser field strength and excitation mechanisms

Multiple ionisation of molecular valence electrons is observed in the $10^{13} - 10^{16}$-W cm^{-2} intensity range. For instance, in the case of a N_2O molecule, a laser pulse of 5×10^{15} W cm^{-2} at $\lambda = 800$ nm is able to remove nine electrons from the molecule and the transient multi-charged molecule $[N_2O^{9+}]$ explodes according to

$$\left[N_2O^{9+} \right] \rightarrow N^{3+} + N^{3+} + O^{3+} + 160 \text{ eV}. \qquad (3.3)$$

The total energy absorbed by the molecule is at least about 460 eV, taking into account the measured release of kinetic energy reported in equation (3.3).

The fundamental mechanism of the laser–molecule coupling is multi-photon absorption. Owing to the complexity of the theoretical multi-photon approach for large numbers of absorbed photons, alternative approaches for the quantitative analysis of the molecular multiple-ionisation process have been developed, e.g. tunnel ionisation [6], numerical integration of the Schrödinger equation [7–9] and Thomas–Fermi models [10]. Several of these theories are detailed in this book.

Recalling that an electric field of 1 a.u. corresponds to a laser intensity of 3×10^{16} W cm^{-2}, the laser–molecule interaction is a highly non-perturbative process in the frame of the zero-field molecule. Following this statement, the electronic field experienced by the nuclei is somewhat different from the electronic field of an unperturbed molecule. In consequence, the nuclear dynamics has to be analysed in detail in order to get more insight into the strong-field effects on the fragmentation processes. In principle, non-perturbative effects take place also during the passage of molecular ions in thin foils in beam–foil experiments, since the inner field in the foil is of the order of 0.1–1 a.u.. However, since the duration of interaction is only a fraction of a femtosecond, the resulting Coulomb explosion gives a picture of the unperturbed molecule as outlined in the previous section. Spectroscopic studies of molecular dications are performed using synchrotron radiation. In these experiments, the one-photon valence-electron excitation remains in the perturbative regime and some similarities to and differences from the laser experiments will be presented. This chapter is limited to valence-electron excitation and, consequently, comparisons with inner-shell excitations are not made.

Multi-charged molecular ions

The dissociation dynamics of a multiply ionised molecule at large internuclear distances is dominated by the Coulomb repulsion of its constituents. However, doubly charged molecules often possess a bound potential well at short internuclear distances due to covalent forces. This well can usually support several vibrational states, which are metastable due to the probability of tunnelling through the Coulomb potential barrier. The lifetimes of these states belong to the 1 ms–1 ps range, depending on their positions in the well. In our experiments, the time of flight of ions is of the order of 1–10 μs and, consequently, a large variety of molecular dications is detected, for instance N_2^{2+}, O_2^{2+}, CO_2^{2+}, $C_2H_n^{2+}$ ($n = 2$ and 4) and $C_3H_n^{2+}$ ($n = 2$ and 4). The triply charged diatomic molecules Cl_2^{3+}, Br_2^{3+} and I_2^{3+} were recently detected using femtosecond laser pulses [11]. When the number of electrons removed increases, the Coulomb repulsion starts to dominate the binding forces and the multiply charged molecule becomes unstable. However, the electronic energy of a multi-charged molecule cannot be described by a simple Coulomb repulsion between the multi-charged ions at short internuclear distances. The contribution of the remaining valence electrons lowers the electronic energy in comparison with the Coulomb energy.

Ion-impact experiments prove that non-Coulombic states play an important role in the molecular explosion and experimental results are confirmed by self-consistent *ab initio* calculations [4, 5]. These observations are possible because the collision time is short in comparison with the timescales of nuclear motions. The measured releases of kinetic energy are compared with the Coulomb-repulsion energies calculated at the molecular equilibrium internuclear distance, since the molecule can be considered frozen during the collision time. The fact that the measured energies are lower than the Coulomb energies signifies the presence of non-Coulombic states. The situation with laser pulses is more difficult since the nuclei have time to move during the laser pulse. In this case, experimental knowledge of the multiple-fragmentation channels is not sufficient to allow one to draw a conclusion about the electronic states of the multi-charged molecules.

3.2 Experimental techniques

The molecular response to strong laser fields can be investigated through the detection of ions, electrons and photons. In this section, only the detection of ions is described in detail since the multiple-fragmentation channels are identified using ion-correlation techniques. The femtosecond-laser systems and the associated diagnostics are discussed elsewhere in this book. The high laser fields are produced using focusing devices such as fused-silica lenses and parabolic mirrors. Owing to the temporal and spatial distributions of laser intensity, the laser intensities reported in this chapter are the peak laser intensities at the best focus.

The ion time-of-flight mass spectrometer

Ion kinematics. The atomic and molecular ions are identified using a set of parameters, $\{M, Z, \boldsymbol{P}\}$, which are the mass, the charge state and the initial momentum of the ion. Conventional ion time-of-flight spectrometry allows us to measure the ratio M/Z with a good accuracy. In the general case, the momentum dependence of the time of flight of a particular ion is a complex function of the spectrometer dimensions and the applied voltages. In 1955, Wiley and McLaren proposed a three-chamber spectrometer in order to introduce a linear dependence on the initial momentum into the time of flight [12]. Figure 3.1 illustrates a spectrometer of this type which is being used in our experiments. The ions are produced in a first zone with a homogeneous electric field and subsequently accelerated further in a second zone before entering the field-free drift tube. The transverse dimensions of the collection and acceleration zones are much larger than the longitudinal dimensions in order to ensure that one has planar potential surfaces in the ion-trajectory space.

Within the space-focusing condition of Wiley and McLaren [12], the time of flight of ions $T(M, Z, \boldsymbol{P})$ can be approximated to a very good accuracy as follows:

$$T(M, Z, \boldsymbol{P}) = T(M, Z, \boldsymbol{P} = 0) - \frac{P\cos(\vartheta)}{ZeF_{\rm c}}, \qquad (3.4)$$

where e is the elementary charge and $F_{\rm c}$ is the collection electric field. In equation (3.4), P is the modulus of the initial momentum $\boldsymbol{P}$ and ϑ is the

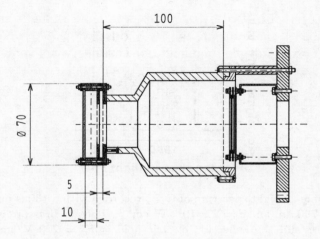

Fig. 3.1. An ion time-of-flight mass spectrometer. The ions are produced in the left-hand zone (10 mm), then are accelerated in the acceleration zone (5 mm) before entering the 100-mm-long drift tube and being detected by the 40-mm-effective-diameter microchannel-plate detector.

angle between P and the axis of the spectrometer directed towards the detector. The space-focusing condition is a precise relationship between the electric fields in the collection and accelerating zones and depends only on the three longitudinal dimensions of the spectrometer. The relative error $\Delta T/T$ between the exact value of the time of flight and the result of equation (3.4) is

$$\frac{\Delta T}{T} = \alpha_s \left(\frac{E \cos^2(\vartheta)}{ZeU_c} \right)^2, \tag{3.5}$$

where α_s depends on the longitudinal dimensions of the spectrometer ($\alpha_s = -5.5 \times 10^{-3}$ for the dimensions given in figure 3.1), $E = P^2/(2M)$ is the initial energy of the ion and ZeU_c is the energy communicated to the ion by the collection electric field. In equation (3.5), $\Delta T/T$ is calculated to lowest order in $E \cos^2(\vartheta)/(ZeU_c)$, which must be smaller than 1 if one is to detect all the ions with initial kinetic energies $E < ZeU_c$. In the present experimental situations, the ratio $E/(ZeU_c)$ is around 0.1 and the relative error introduced by equation (3.4) remains, therefore, very small.

Figure 3.2 represents a time-of-flight spectrum of C_2H_2 that was recorded with a laser pulse of duration 130 fs. Single peaks are ascribed to the singly and doubly charged molecular ions $C_2H_2^+$ and $C_2H_2^{2+}$. The hydrogen ions H^+ and the multi-charged carbon ions C^{Z+} ($Z = 1$, 2 and 3) originate from the Coulomb explosion of the molecule and display a highly symmetric double-peak structure due to the operating condition given by equation (3.4). The double-peak structures in figure 3.2 are

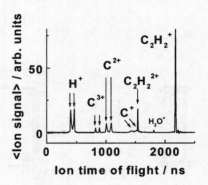

Fig. 3.2. Time-of-flight spectrum of C_2H_2 recorded with a 130 fs pulse duration laser at $\lambda = 790$ nm and $I = 2 \times 10^{15}$ W cm^{-2}. The gas pressure is $p(C_2H_2) = 4 \times 10^{-9}$ Torr and the collection electric field is $F_c = 200$ V/cm. The laser polarisation direction is parallel to the spectrometer axis.

related to the fact that the linear C_2H_2 molecule is aligned along the axis of the spectrometer by the intense laser field. Each class of ions is thus composed of ions ejected towards the detector (*forward* ions with $\cos(\vartheta) > 0$) and ions ejected in the opposite direction (*backward* ions with $\cos(\vartheta) < 0$). On applying the linear frame transformation of equation (3.4), the time-of-flight spectrum gives the momentum spectrum of the H^+ and C^{Z+} ($Z = 1, 2$ and 3) ions.

Angular discrimination. Following figure 3.1, the ions are accelerated along the axis of the spectrometer, while their drift velocity perpendicular to this direction remains unchanged. If the time of flight is too long, ions with non-negligible perpendicular momenta can miss the detector at the exit of the drift tube. In order to avoid this angular discrimination, the first possibility is to increase the collection electric field. However, according to equation (3.4), the momentum resolution will be poor since the electric field F_c appears in the denominator. A second solution is to increase the dimensions of the detector, but microchannel-plate detectors are quite expensive for effective diameters above 40 mm. One of the possible solutions is to work with short drift tubes and to adjust the collection electric field to collect all the ions coming from the molecular explosion. For the dimensions given in figure 3.1, an ion ejected perpendicularly to the axis of the spectrometer is detected on the 40-mm microchannel-plate detector if its initial energy E satisfies the condition

$$E < 0.122 Z e U_c, \tag{3.6}$$

where ZeU_c is defined as previously. The potential U_c or, equivalently, the collection field F_c is set according to equation (3.6) in order to collect all ions. This point is particularly important for multiple fragmentation, in which case the atomic ions can be ejected in any direction.

Vacuum. Laser-induced multiple ionisation is a non-resonant process and all the molecular species present in the interaction volume are ionised. In consequence, a high-vacuum system is required in order to reduce the contribution of the background gas to the ion signal. Using a single $400 \, \mathrm{l \, s^{-1}}$ turbopump and a vessel adapted to the spectrometer of figure 3.1, a base pressure of the order of 10^{-10} Torr is achieved. In addition, as shown by figure 3.2, a very low operating gas pressure produces sufficient signal. Moreover, too high a gas pressure can give rise to space-charge effects, which severely perturb the operating condition of the spectrometer given in equation (3.4). Another important reason for working with low gas pressures is given in the next section and is necessary for the success of ion-correlation techniques.

The identification of multiple-fragmentation channels

Information on the transient precursor $[ABC^{M+}]$ of a multiple-fragmentation process, for instance the case

$$[ABC^{M+}] \rightarrow A^{Z+} + B^{Z'+} + C^{Z''+} + \text{Energy}, \qquad (3.7)$$

comes only from the fragments. It is straightforward to identify the parameters $\{i\} = \{M, \, Z, \, \boldsymbol{P}\}$ for every fragment ion in the time-of-flight spectrum, but this is not sufficient for the determination of the fragmentation channels. In practice, the fragment signals $s(i)$ of a large number of multiple-fragmentation events are recorded and the experimentalist is presented with a large set of data $\{s_k(i), \, k = 1, \ldots, N\}$, where N is the total number of laser shots. Depending on the average number of multiple-fragmentation events per laser shot, the ion correlations are extracted by two different methods. The first one is the multiple-coincidence technique, which pertains under the condition of there being less than one multiple-fragmentation event per laser shot. The second method is the statistical multiple-correlation technique, introduced by the Reading group, which tolerates more than one event per laser shot [13].

Multiple-coincidence techniques. When several fragments are detected within a short timespan, they originate almost certainly from the same dissociation process, if the probability of two molecules fragmenting during that time window is negligibly small. In this case, the ion signals $s_k(i)$ are transformed into binary values $\{0, 1\}$ by a fast discriminator. When one is using 'unpulsed' excitation sources such as synchrotron radiation and electron or ion impact, the electronic devices for the ion analysis

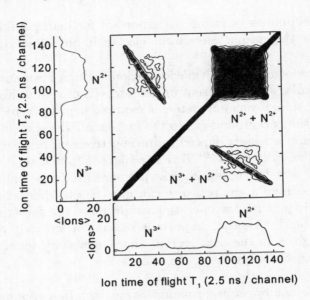

Fig. 3.3. An ion–ion coincidence map recorded with a 40-fs circularly polarised laser pulse at $\lambda = 800$ nm, $I = 4 \times 10^{15}$ W cm^{-2} and $p(N_2) = 2 \times 10^{-9}$ Torr. The collection electric field was $F_c = 180$ V cm^{-1}. The bottom and left-hand curves represent the average ion signal as a function of the times of flight T_1 and T_2. The ion–ion coincidence signal is represented by the central map using a five-level grey scale as a function of the T_1 (horizontal axis) and T_2 (vertical axis) flight times of ions.

(time-to-digital converters) require a *START* signal, which could be, for example, an electron signal coming from the initial multiple-ionisation process [14, 15]. The ion signals are subsequently recorded as *MULTI-STOP* events and doublets (i, j) and triplets (i, j, k) in the time-of-flight spectrum represent, respectively, two- and three-ion coincidence events. The problem of false-coincidence counting rates is addressed in [16].

In laser experiments, the *START* signal is delivered by a fast photodiode, which detects the laser pulse. Figure 3.3 presents an ion–ion coincidence map recorded for N_2 at $p(N_2) = 2 \times 10^{-9}$ Torr. The experimental data are recorded with analogue signals $s_k(i)$ coming from the microchannel-plate detector. The ion–ion coincidence signal is proportional to

$$C_2^{(p)}(i, j) = \frac{1}{N} \sum_{k=1}^{N} s_k(i) s_k(j). \tag{3.8}$$

The conventional N^{2+} and N^{3+} ion time-of-flight spectra are represented by the bottom and left-hand curve and are the average ion signal $C_1(i) = \sum_{k=1}^{N} s_k(i)/N$. The central map shows the ion–ion coincidence signal given by equation (3.8). Although the map is dominated by true coincidences, false-coincidence signals are noticeable for the $N^{2+} + N^{2+}$ fragmentation channel. The shapes of the coincidence peaks will be discussed in the next section. Using intense lasers, the probability of ionisation is very high and every molecule present in the interaction volume is ionised. As illustrated in figure 3.3, it is possible to use the coincidence techniques with lasers, but the operating pressure must remain very low. However, in most practical situations, the statistical method is an easier technique for the identification of the fragmentation channels. The advent of very short laser pulses of around 5 fs [17] might favour a more extensive use of coincidence methods, since the number of events decreases with the pulse duration.

Statistical multiple-correlation techniques. This technique was introduced in order to overcome the problem of false coincidences in Coulomb-explosion experiments in strong laser fields [13]. The ion signal $\{s_k(i)\}$ coming from the detector is no longer transformed to a binary signal by a fast discriminator, but rather its analogue nature is retained in order to take account of the fluctuations in number of ions in channel i for each laser shot k. These fluctuations arise from the probabilistic nature of the fragmentation process, which can follow various fragmentation pathways. In order to determine whether the ion signals in channels i and j are correlated, i.e. whether the corresponding ions come from the same fragmentation channel, one has to apply a two-entry statistical coefficient called the covariance coefficient. The ion–ion coincidence coefficient in equation 3.8 is replaced by

$$C_2(i,j) = \langle\, [s(i) - \langle s(i)\rangle]\,[s(j) - \langle s(j)\rangle]\,\rangle, \tag{3.9}$$
$$= \langle s(i)s(j)\rangle - \langle s(i)\rangle\langle s(j)\rangle,$$
$$= \frac{1}{N}\sum_{k=1}^{N} s_k(i)s_k(j) - \left(\frac{1}{N}\sum_{k=1}^{N} s_k(i)\right)\left(\frac{1}{N}\sum_{k=1}^{N} s_k(j)\right). \tag{3.10}$$

The notations $\langle\ \rangle$ represent the average over N laser shots. The extension of equation (3.9) for a three-entry covariance coefficient $C_3(i,j,k)$ is straightforward and is used to study three-body fragmentation channels, for instance for triatomic molecules.

The statistical method can be extended without any conceptual difficulties to signals coming from particles detected with various devices, such as electrons and ions, provided that these particles are produced during the same excitation event. In the case of ion–ion correlations, the same signals $s_k(i)$ appear as inputs both in $C_2(i,j)$ and in $C_3(i,j,k)$ and these coefficients are therefore symmetric with respect to (i,j,k) permutations.

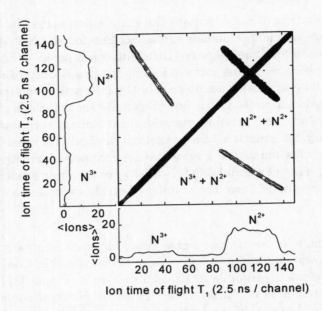

Fig. 3.4. An ion–ion covariance map recorded with a 40-fs circularly polarised laser pulse at $\lambda = 800$ nm, $I = 4 \times 10^{15}$ W cm^{-2} and $p(N_2) = 2 \times 10^{-9}$ Torr. The collection electric field is $F_c = 180$ V cm^{-1}. The bottom and left-hand curves represent the average ion signal as a function of the times of flight T_1 and T_2. The ion–ion covariance coefficient is represented by the central map using a five-level grey scale as a function of the T_1 (horizontal axis) and T_2 (vertical axis) times of flight of ions.

When $i = j$ in equation (3.9), the coefficient $C_2(i,i)$ is the variance of the ion signal. Equivalently, the coefficient $C_1(i) = \langle s(i) \rangle$ is the average ion signal and gives the conventional ion time-of-flight spectrum. The first term in equation (3.10) is proportional to the ion–ion coincidence coefficient, whilst the effect of the second term is to remove the contribution from false coincidences. However, this picture has to be considered with care, because the ion signals no longer belong to the binary world of coincidence experiments.

Figure 3.4 presents the covariance coefficient of the data for which figure 3.3 showed the ion–ion coincidence coefficient. Now the correlation islands display the correct behaviour, i.e. without a significant contribution from false correlations. The correlation islands associated with the channels $N^{Z_1+} + N^{Z_2+}$ are located on lines $Z_1 T_1 + Z_2 T_2 = constant$, due to a combination of conservation of momentum, $\boldsymbol{P}_1 + \boldsymbol{P}_2 = 0$, and the time–momentum relationship given by equation (3.4).

Several conditions are necessary for the correct recording of statistical correlation maps. The first one is that it is necessary to work with a sufficiently low gas pressure in order to get a linear response of the detector as a function of the number of incident impinging ions. The second one is related to the non-linear nature of the laser–molecule coupling. A poorly stable laser will introduce large fluctuations into the ion signals, which will dominate the statistics of the overall detection. The ion-spectrometer requirements are the same as for coincidence experiments, i.e. high-transmission grids and a good detector response are essential. The minimum number of laser shots that is required for the identification of the fragmentation channels depends on their branching ratios. Typical numbers are of the order of 10^4 for double correlations and 10^5 for triple correlations. Finally, this method does not allow a straightforward detection of fragmentation channels involving neutral fragments. However, a fragment ion in the time-of-flight spectrum that is not correlated to any island on the covariance map indicates the presence of one or more undetected neutral companions.

3.3 Two-charge separation fragmentation

The charge separation resulting from the multiple ionisation of diatomic molecules was discussed in chapter 2. The experimental identification of the fragmentation channels is based on a two-entry covariance analysis since only two atomic ions are involved in the fragmentation. In larger molecules, double- and multiple-ionisation processes can lead to a wide variety of fragmentation processes. The triatomic and larger molecules are discussed separately in this section because it is possible to investigate the whole atomisation of a triatomic molecule in detail and relate this process to two-charge separation channels. For larger systems, however, the correlation techniques become difficult to use. This problem will be addressed in the next section. Two-charge separation fragmentations that involve a neutral fragment are not included, although they are currently being observed in our experiments.

Triatomic molecules

The charge separation of a triatomic system involves an atomic ion and a molecular ion: $[ABC^{M+}] \rightarrow A^{Z+} + BC^{(M-Z)+}$. The first point is to investigate the dynamics of the fragmentation. For instance, double ionisation of neutral molecules, which leads to the detection of a molecular dication as well as two-charge separation channels, is a non-sequential electron-emission process at low laser intensities [18]. Another interesting aspect is the comparison of the laser-excitation results with those obtained with other sources of valence-electron excitation, for example synchrotron radiation in the range 35–65 eV and double-charge-transfer excitation

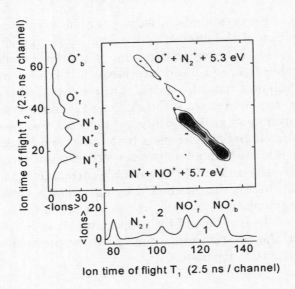

Fig. 3.5. An ion–ion covariance map recorded with a 40-fs linearly polarised laser pulse at $\lambda = 800$ nm, $I = 5 \times 10^{14}$ W cm^{-2} and $p(N_2O) = 10^{-8}$ Torr. The collection electric field is $F_c = 500$ V cm^{-1}. The bottom and left-hand curves represent the average ion signal as a function of the times of flight T_1 and T_2. The ion–ion covariance coefficient is represented by the central map using a five-level grey scale as a function of the T_1 (horizontal axis) and T_2 (vertical axis) times of flight of ions.

[19, 20].

Figure 3.5 displays the ion–ion covariance map for the identification of the fragmentation channels that result from the double ionisation of the N_2O molecules. The bottom time-of-flight spectrum shows the forward NO_f^+ and backward NO_b^+ components of the NO^+ ion, whilst the corresponding components of the N_2^+ ion are very weak. The associated atomic ions appear in the left-hand time-of-flight spectrum. In particular, the N^+ ions exhibit a three-peak structure. The forward N_f^+ and backward N_b^+ originate from the terminal N atom in N_2O, whilst the central N_c^+ component comes from the central N atom. In figure 3.5, two dissociation channels are detected:

$$[N_2O^{2+}] \rightarrow N^+ + NO^+ + 5.7 \text{ eV}, \tag{3.11}$$

$$[N_2O^{2+}] \rightarrow N_2^+ + O^+ + 5.3 \text{ eV}. \tag{3.12}$$

Process (3.11) largely dominates over process (3.12): the relative proportions of the covariance coefficients are 8 : 1. The ion–ion covariance islands are found on a line of slope −1, because of the conservation of momentum

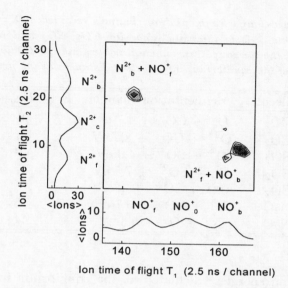

Fig. 3.6. An ion–ion covariance map recorded at $\lambda = 400$ nm and $p(N_2O) = 2.2 \times 10^{-8}$ Torr for identification of the $N^{2+} + NO^{+} + 9.5$ eV channel. The collection electric field is $F_c = 500$ V cm^{-1}.

$P_1 + P_2 = 0$. Finally, the central peaks labelled 1 and 2 in the bottom time-of-flight spectrum correspond, respectively, to zero-momentum NO^{+} and N_2^{+} ions and do not give rise to any correlation to atomic ions because their companion fragment is an undetected neutral N or O atom.

Two-charge separation channels involving doubly charged atomic and/ or molecular ions have also been detected in laser-induced multiple ionisation [21]. For the sake of simplicity, we remain with the same molecular species and show in figure 3.6 the $N^{2+} + NO^{+} + 9.5$ eV channel originating from the triple ionisation of N_2O at $\lambda = 400$ nm. In this figure, the pattern of the N^{2+} peaks is the same as the pattern of the N^{+} peaks in figure 3.5. Neither the zero-momentum NO_0^{+} ion nor the central N_c^{2+} ion gives any correlation signal, as expected. The covariance islands are found on a line of slope -2, because of conservation of momentum.

A selection of measured two-charge separation channels at $\lambda = 800$ nm is listed in table 3.1 for CO_2, SO_2 and N_2O. These results have been obtained using laser-pulse durations in the range 40–130 fs and do not exhibit any noticeable variation with pulse duration, just as in the case of diatomic molecules [22]. Moreover, some of these channels are also observed at $\lambda = 400$ nm with practically the same fragmentation energies. Once more we note a connection with results previously obtained with diatomic molecules.

The fragmentation energies listed in table 3.1 are lower than the Cou-

Table 3.1. *Two-charge fragmentation channels recorded at $\lambda = 800$ nm in the 10^{14}–10^{15} W cm^{-2} laser-intensity range for CO_2, SO_2 and N_2O. The ratios E_{Exp}/E_{Coul} of the measured experimental values and the Coulomb energies are calculated using the equilibrium internuclear distances of the neutral molecules*

Molecule	Fragmentation channel	E_{Exp}/E_{Coul} (%)
CO_2	$O^+ + CO^+ + 4.7$ eV	38
CO_2	$O^{2+} + CO^+ + 8.5$ eV	34
SO_2	$O^+ + SO^+ + 4.0$ eV	40
SO_2	$O^+ + SO^{2+} + 7.5$ eV	37
SO_2	$O^{2+} + SO^{2+} + 18$ eV	45
N_2O	$N^+ + NO^+ + 5.7$ eV	45
N_2O	$N_2^+ + O^+ + 5.3$ eV	44

lomb repulsion energies calculated at the equilibrium internuclear distances. In particular, for channels resulting from the double ionisation of molecules, the energies are in the same range as those measured with other means of excitation of valence electrons. For example, for N_2O the kinetic energies and branching ratios are very close to the results of Price *et al.*, which were obtained from double-charge-transfer and photo-ionisation measurements [20]. Interestingly, the dominant channel $N^+ + NO^+ + 5.7$ eV is counter-intuitive from the following consideration. The dissociation energies of the neutral molecule, $D(N–NO) = 4.930$ eV and $D(N_2–O) = 1.677$ eV, suggest that it is easier to break the molecule via the $N_2 + O$ channel. On the other hand, the dissociation pathways of the molecular dication differ strongly from those of the neutral molecule. Using *ab initio* calculations of potential-energy surfaces of the N_2O^{2+} dication, Levasseur and Millié [23] found that the dissociation into $N^+ + NO^+$ is a direct process, whilst high and wide potential barriers considerably inhibit the dissociation into $N_2^+ + O^+$. This interpretation is confirmed here by laser experiments. However, the similarity of the response of N_2O to different excitation sources is not a general rule, as we will see for C_2H_2.

Molecules with more than three atoms

A set of experiments was performed with molecules built with carbon and hydrogen atoms. The presence or absence of π electrons in certain molecular species allows us to gain some insight into the effects of the electronic configuration on the fragmentation processes. For instance, figure 3.2 displays the $C_2H_2^+$ and $C_2H_2^{2+}$ molecular ions as well as the H^+ and C^{Z+} ($Z = 1$, 2 and 3) atomic ions. However, CH^+, CH_2^+ and

C_2H^+ ions are not detected. Consequently, two-body reactions, such as $CH^+ + CH^+$, $C^+ + CH_2^+$ and $H^+ + C_2H^+$ do not occur. On the other hand, photo-ionisation of C_2H_2 in the photon-energy range between 35 and 65 eV does produce all these dissociation pathways [19]. Thus, in contrast to the case of N_2O, the C_2H_2 molecule responds quite differently to intense-laser and single-photon excitations of valence electrons.

Whilst the single-photon ionisation results can be understood from the electronic potential surfaces of the $C_2H_2^{2+}$ dication, with laser excitation, the laser field couples strongly to all four π electrons of the neutral molecule. In contrast to the case of N_2O, the stripping away of electrons is therefore too fast to allow a two-charge separation in C_2H_2. Similar behaviour is also found for the C_2H_4 molecule in the range 10^{14}–10^{15} W cm^{-2}. Again, CH_n^+ ions are not detected, even though the molecule has only a π^2 configuration. Nevertheless, small peaks of $C_2H_3^+$ and $C_2H_2^+$ ions are observed in ratios of less than $1:10$ with the main $C_2H_4^+$ peak. Finally, the bonds of the C_3H_6 molecule are saturated with σ electrons. Consequently, when these electrons are removed from the molecule, a strong two-charge separation channel,

$$[C_2H_6^{2+}] \rightarrow CH_3^+ + CH_3^+ + 4.0 \text{ eV}, \tag{3.13}$$

is observed in addition to the many-body fragmentation processes.

Another example of the effect of the electronic configuration is given in figure 3.7, which presents the mass spectra of allene and propane. In the spectrum of allene, the $C_2H_n^+$ contributions are very small in comparison with those of the corresponding ions in the spectrum of propane. For allene, two-charge separation channels such as $C_3H_3^+ + H^+$ are detected but the C_3 chain of the molecule does not undergo any significant two-body fragmentation. The Coulomb explosion of the C_3 structure is therefore an instantaneous three-body fragmentation. Since the carbon skeleton of the molecule remains bound by σ electrons, the $C_3H_n^{2+}$ ($n = 2$ and 4) dications are stable species within the time-of-flight window and can be detected. This overall behaviour is not observed for the C_3H_8 molecule because the ejection of σ bonding electrons favours the two-charge separation channels in addition to the three-body fragmentation of the carbon chain [24].

3.4 Full-atomisation fragmentation

In the previous section, many two-body charge separation channels were reported. In this section, it is shown that the three-body fragmentation channels of a triatomic molecule originate from a direct rupture of the two bonds and are independent of the two-body fragmentation channels.

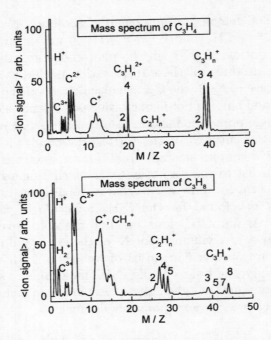

Fig. 3.7.   Mass spectra of C_3H_4 and C_3H_8 recorded at $\lambda = 790$ nm with a 130-fs pulse at $I = 2.5 \times 10^{15}$ W cm^{-2} and $p = 10^{-9}$ Torr. The ratio M/Z is the mass divided by the charge of the detected ions.

The non-sequential, four-body fragmentation of C_2H_2 is also commented on, as an example for larger molecular systems.

Sequential versus non-sequential fragmentation

When three multiply charged atomic ions are detected in coincidence following the multiple ionisation of a triatomic molecule, the main question is that of whether the rupturing of bonds is a sequential or a non-sequential process, as illustrated in figure 3.8. In the sequential process, the molecule initially dissociates via two-body fragmentation. Following the notation in figure 3.8, this first step gives rise to a single atom or ion (1) with momentum $\boldsymbol{P}_1$ and a diatomic, charged or neutral molecule (23) with opposite momentum, i.e. $\boldsymbol{P}_{23} = -\boldsymbol{P}_1$. While atom (1) maintains its momentum, the molecular system (23) could undergo a uni-molecular fragmentation and even interact further with the laser pulse. The resulting fragmentation into atoms (2) and (3) occurs in the drifting molecular frame and obeys conservation of momentum, $\boldsymbol{P}'_2 + \boldsymbol{P}'_3 = 0$. The momenta in the laboratory frame are subsequently found from the Galilean velocity-frame transformation.

In the corresponding non-sequential fragmentation, no intermediate

SEQUENTIAL THREE-BODY FRAGMENTATION

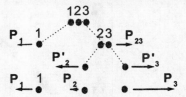

NON-SEQUENTIAL THREE-BODY FRAGMENTATION

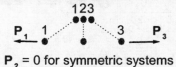

$P_2 = 0$ for symmetric systems

Momentum detection $\{P_1, P_2, P_3\}$
of particles 1, 2, and 3

Fig. 3.8. Sequential versus non-sequential fragmentation for a triatomic molecule composed of atoms labelled 1, 2 and 3.

two-body fragmentation occurs, as illustrated in figure 3.8. A precise definition of a direct or non-sequential process is in fact difficult to establish. Indeed, figure 3.8 represents the two limiting cases of fragmentation behaviour. In between, various physical situations might occur, e.g. a stretching of one bond, followed by a rapid dissociation of the other. Possibly, a definition of non-sequential multiple fragmentation could be given in the time domain by requiring that the full atomisation should be completed within one vibrational period. Another possible definition could be in the space domain: the three particles are ejected individually out of a box of the molecular size of about 5 Å. However, the experimental apparatus cannot provide such temporal and spatial resolutions. We therefore use a third criterion that relates to momentum space: the non-sequential multiple fragmentation of a symmetric system (e.g. $O^{Z_1+} + C^{Z_2+} + O^{Z_1+}$ from CO_2) is equivalent to the observation of a zero-momentum central particle (here C^{Z_2+}).

Non-sequential multiple fragmentation of CO_2, N_2O and C_2H_2

Figure 3.9 is a covariance map recorded with the C^{2+}, C^{3+}, O^{2+} and O^{3+} ions for the identification of the $O^{Z_1+} + C^{Z_2+} + O^{Z_3+}$ $(Z_1, Z_2, Z_3 = 2, 3)$ channels. The oxygen forward and backward components appear clearly in the time-of-flight spectrum, while the central carbon atom produces single C^{2+} and C^{3+} ion peaks. $C_3(T_1, T_2, T_3)$ is a three-entry coefficient and thus requires three coordinate axes for its representation. For a more

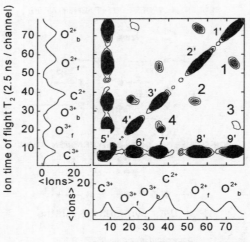

Fig. 3.9.   A three-entry covariance map of CO_2 recorded at $\lambda = 800$ nm, $I = 4 \times 10^{15}$ W cm^{-2} and $p(CO_2) = 10^{-9}$ Torr. The laser-polarisation direction is parallel to the axis of the ion spectrometer and the collection electric field is $F_c = 500$ V cm^{-1}. The coefficient $C_3(T_1, T_2, T_3)$ is represented by the central map as a function of (T_1, T_2) with a fixed value $T_3 = T_3^0$, which corresponds to zero-momentum C^{3+} ions. The islands labelled $1'$–$9'$ correspond to partially autocorrelated signals with $T_1 = T_2$ ($1'$–$5'$) and $T_2 = T_3^0$ ($5'$–$9'$). The three-ion channels are labelled as follows: island 1, $O_f^{2+} + C^{3+} + O_b^{2+}$; island 2, $O_b^{3+} + C^{3+} + O_f^{2+}$; island 3, $O_f^{3+} + C^{3+} + O_b^{2+}$; and island 4, $O_f^{3+} + C^{3+} + O_b^{3+}$.

convenient two-dimensional analysis and representation, the time of flight T_3 is fixed to $T_3^{(0)}$ and the resulting coefficient $C_3(T_1, T_2, T_3 = T_3^{(0)})$ is analysed as a function of $T_3^{(0)}$. For CO_2, this coefficient exhibits maxima for zero-momentum C^{Z+} ions from symmetric channels. Figure 3.9 represents the covariance coefficients for the value $T_3^{(0)}$ that corresponds to a zero-momentum C^{3+} ion. Islands 1–4 give a clear identification of the corresponding three-body fragmentation channels, whilst islands $1'$–$9'$ are due to partially autocorrelated signals, i.e. when two times of flight (T_i and T_j) in the coefficient $C_3(T_1, T_2, T_3)$ are nearly the same.

There are several ways of choosing the value $T_3 = T_3^0$. For instance, figure 3.10 shows the three-entry covariance map for the identification of the $N_f^{3+} + N_c^{3+} + O_b^{3+} + 160$ eV channel. In this case, the value T_3^0 is scanned for the O_b^{3+} ion species. In figure 3.10, the forward and central components of the N^{3+} ion are correlated to the O_b^{3+} ion. Moreover, the backward N_b^{3+} ion does not give any correlation signal, in accordance with simple kinematics arguments. As in the case of CO_2, the central nitrogen

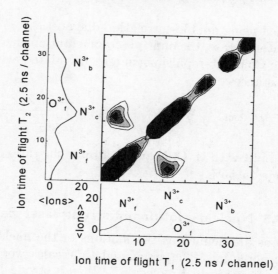

Fig. 3.10. A three-entry covariance map of N_2O recorded at $\lambda = 800$ nm, $I = 6.5 \times 10^{15}$ W cm^{-2} and p(N_2O) $= 3 \times 10^{-9}$ Torr. The laser-polarisation direction is parallel to the axis of the ion spectrometer, and the collection electric field is $F_c = 400$ V cm^{-1}. The coefficient $C_3(T_1, T_2, T_3)$ is represented on the central map as a function of (T_1, T_2) with a fixed value $T_3 = T_3^0$, which corresponds to O_b^{3+} ions. The channel $N_f^{3+} + N_c^{3+} + O_b^{3+} + 160$ eV is identified from the off-diagonal islands. As usual, the diagonal patterns originate from partially autocorrelated ion signals.

ion does not exhibit any significant release of kinetic energy, even though the N_2O molecule is not symmetric. This experimental result is confirmed by the ion energies calculated assuming a pure Coulomb explosion.

These two examples, CO_2 and N_2O, show that full atomisation is a non-sequential, three-body, multiple-fragmentation process. In other words, the two-body fragmentation channels reported in table 3.1 are not correlated to the three-body fragmentation channels. All these channels appear as independent decay channels following the laser excitation. The non-sequential behaviour is also observed for the C_3 chain of C_3H_3, with basically the same characteristics as in the case of CO_2 [24, 25]. Because of the increasing complexity of the correlation technique with the number of particles, no trials were undertaken with a four-entry coefficient $C_4(T_1, T_2, T_3, T_4)$ to identify, for instance, the four-body fragmentation of C_2H_2. However, using the two-entry covariance coefficient $C_2(T_1, T_2)$, it is possible to show that the multiple fragmentation in C_2H_2,

$$[C_2H_2^{M+}] \rightarrow H^+ + C^{Z+} + C^{Z'+} + H^+ \qquad (Z = 1, 2, 3), \qquad (3.14)$$

is a non-sequential process. This behaviour is almost clear in figure 3.2

and is confirmed using double-correlation diagnostics.

As in the diatomic case, the multiple-fragmentation energies are specific fractions of the Coulomb-repulsion energies calculated at the equilibrium internuclear distances:

$$E_{\text{Exp}} = \alpha E_{\text{Coul}} \qquad \text{where} \qquad E_{\text{Coul}} = \sum_{j<i} \frac{Z_j Z_i}{|\boldsymbol{R}_j^{\text{e}} - \boldsymbol{R}_i^{\text{e}}|} \qquad (3.15)$$

For molecules built with H, C, N and O atoms, the ratio $\alpha = E_{\text{Exp}}/E_{\text{Coul}}$ lies in the range 45–55%.

3.5 Nuclear motions in strong laser fields

Nuclear motions are defined as the motions of the nuclei in the field of the electronic potential. The associated timescales were discussed previously for the unperturbed molecule. When a strong laser field is applied to a molecular system, the molecular electrons experience oscillations in the field that strongly perturb the zero-field electronic potential. For instance, the mean kinetic energy of a free electron in an electric field $F(t) = F_0 \cos(\omega t)$ is the well-known ponderomotive potential $U_{\text{p}} = q_{\text{e}} F_0^2/(4 m_{\text{e}} \omega^2)$, where m_{e} and q_{e} are the electron's mass and charge. For a laser intensity and wavelength pertaining to current molecular multiple-ionisation experiments, i.e. $I = 10^{15}$ W cm^{-2} and $\lambda = 800$ nm, the ponderomotive energy is equal to $U_{\text{p}} = 60$ eV. Although the bound and departing electrons cannot be considered free particles, the value of U_{p} gives an order of magnitude for the perturbation by the laser field. Following the molecular classification of nuclear motions, the discussion is split into processes of molecular rotational alignment and vibrational motions.

Molecular alignment

In the absence of a laser field, the molecular rotation is a free motion. For the simplest systems, the rotational energy E_{Rot} is quantised according to $E_{\text{Rot}}(J) = B_{\text{e}} J(J+1)$ to a first approximation, where B_{e} is the rotational constant. The values of B_{e} (in meV) are listed in table 3.2 for H_2, N_2, N_2O and CO_2. In multiple ionisation, a large electronic polarisability is induced by the intense laser field. Recent experiments show that light molecular systems such as H_2 and N_2 are reoriented along the direction of polarisation of the laser, even with laser-pulse durations as short as 50 fs [26–28]. Various approaches can be used to describe this phenomenon. We propose to discuss the alignment of linear molecular systems in terms of the polarisability tensor α_{ij} of the neutral molecule. Although the components of this tensor depend on the photon energy $h\nu$,

Table 3.2. *The zero-field rotational constant B_e, maximum well depth ΔV_{Max}, and rotation time T_{Rot} calculated for a peak laser intensity $I_{Max} = 10^{15}$ W cm^{-2}*

Molecule	B_e (meV)	ΔV_{Max} (meV)	T_{Rot} (fs)
H_2	3.888	21.9	160
N_2	0.248	96.8	422
N_2O	0.052	290	531
CO_2	0.048	212	645

they can nevertheless be replaced by the static components of α_{ij}, since $h\nu$ remains small in comparison with the lowest electronic excitation energies. Whilst this model cannot explain the molecular reorientation in detail, since it does not take into account the whole of the dynamics of multiple ionisation, it can still give a good qualitative description of what happens in molecules.

To first order, the laser field $\boldsymbol{F}$ induces a diplole $\boldsymbol{p}$ with components $p_i = \varepsilon_0 \sum_j \alpha_{ij} F_j$. In a linearly polarised laser field, the time-dependent potential experienced by the nuclei, averaged over one laser period, $V(t, \vartheta) = -\frac{1}{2}\langle \boldsymbol{p} \cdot \boldsymbol{F} \rangle$ is given by

$$V(t, \vartheta) = -\frac{I(t)}{2c} \left[(\alpha_{\parallel} - \alpha_{\perp}) \cos^2(\vartheta) + \alpha_{\perp} \right], \qquad (3.16)$$

where $I(t)$ is the time-dependent laser intensity and c stands for the speed of light. In equation (3.16), the angle ϑ is the angle between the laser's electric field and internuclear axis and $\alpha_{\parallel}$ and $\alpha_{\perp}$ are the parallel and transverse components of the polarisability tensor. The difference $\Delta\alpha = \alpha_{\parallel} - \alpha_{\perp}$ represents the anisotropy of the polarisability tensor and is responsible for the well structure of $V(t, \vartheta)$ as a function of ϑ. The maximum well depths $\Delta V_{Max} = I_{Max} \Delta\alpha/(2c)$, which occur at $\vartheta = 0$ and the peak laser intensity $I_{Max} = 10^{15}$ W cm^{-2}, are listed in table 3.2, for comparison with the B_e values for H_2, N_2, N_2O and CO_2.

In table 3.2, the maximum well depths are larger than the field-free rotational constants B_e and, consequently, the molecular rotation is no longer a free motion in the high laser fields used in the molecular multiple-ionisation experiments. Using the classical equation of motion $J \, d^2\vartheta/dt^2 = -\partial V/\partial\vartheta$, where J is the molecular momentum of inertia, the temporal rotational behaviour is expressed for the angle $\Theta = 2\vartheta$ as follows:

$$\frac{d^2\Theta}{dt^2} + \left(\frac{2\pi}{T_{Rot}} \right)^2 f(t) \sin(\Theta) = 0, \qquad (3.17)$$

where $f(t)$ is the temporal envelope of the laser intensity $I(t) = I_{Max} f(t)$. The time constant T_{Rot} gives an order of magnitude for the molecular rotation:

$$T_{\text{Rot}} = 2\pi \sqrt{\frac{cJ}{I_{\text{Max}} \Delta \alpha}}. \tag{3.18}$$

The values of T_{Rot} reported in table 3.2 are much larger than the laser periods for wavelengths in the visible range, and justify *a posteriori* the average of the laser–molecule interaction $V(t, \vartheta) = -\frac{1}{2}\langle \boldsymbol{p} \cdot \boldsymbol{F} \rangle$. However, for light molecules and in particular for H_2, these values cannot explain the observed molecular reorientation for a pulse duration of 50 fs [26, 27]. Of course, the multiple ionisation is not taken into account in this simple model. Since the electronic dynamics occurs basically along the laser-polarisation direction, the departing electrons possibly induce additional forces and, therefore, torques on the nuclear arrangement. Thus the actual rotation time will be shorter than that given by equation (3.18), which is based on the polarisability of the neutral molecule. For the molecules in table 3.2, which are built with light atomic elements, molecular reorientation is observed for pulse durations as short as 40 fs at laser intensities above $I_{\text{Max}} = 10^{15}$ W cm^{-2}. According to equation (3.18), the rotation time should be longer for lower laser intensities. Indeed, in figure 3.5, for which $I = 5 \times 10^{14}$ W cm^{-2}, there is still a covariance signal for ions with zero momentum along the axis of the spectrometer, meaning that there still remain some non-oriented molecules. Considering the angular distributions of the ejected fragments, we expect from equation (3.18) that their widths will increase for lower laser fields. In addition, the rotation dynamics should depend on the time within the laser pulse, during which the multiple ionisation occurs.

Bending and stretching

Prior to laser excitation, neutral molecules will generally be in their ground vibrational state. The nuclear positions are distributed according to the square of the vibrational wavefunction, $|\psi_0(q)|^2$, where q is the vibrational mode coordinate (length or angle):

$$|\psi_0(q)|^2 = \frac{1}{\sigma\sqrt{\pi}} \exp\left[-\left(\frac{q-q_0}{\sigma}\right)^2\right], \tag{3.19}$$

where q_0 is the equilibrium coordinate and $\sigma = [\hbar/(\mu_q \omega)]^{\frac{1}{2}}$ is a measure of the extension of the probability density as a function of the mode pulsation ω and the generalised mass μ_q associated with the coordinate q.

Whilst the rotational motion is probed by a direct position-sensitive method, the vibrational motions must be deduced from a comparison between measured fragmentation patterns and modelled patterns in which

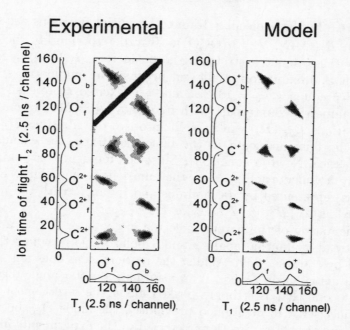

Fig. 3.11. Experimental and calculated ion–ion correlation maps for CO_2. The experimental data are recorded at $\lambda = 790$ nm, $I = 10^{15}$ W cm^{-2} and with a pulse duration of 130 fs. The laser polarisation is directed parallel to the axis of the ion spectrometer and the collection electric field is $F_c = 500$ V cm^{-1}.

the nuclear-coordinate distributions are introduced as parameters, as illustrated in figure 3.11. As explained in the experimental section, the ion spectrometer determines only the component of the ion momentum $\{P_i\}$ along its axis. Since the molecular system is oriented along the direction of the laser field, the variation of this direction relative to the axis of the spectrometer allows us to deduce both the parallel and the perpendicular components of the momenta. However, in order to get the initial positions $\{R_i\}$ of the nuclei, it is necessary to introduce into the model the repulsive potentials that will link the measured momenta to the initial positions. In a first approach, we use pure Coulomb potentials. This approximation should be valid for internuclear distances larger than typically 3 Å, for which the potential curves are essentially Coulombic.

In order to compare the experimental and calculated momentum spectra, several frame transformations are necessary. The Coulomb explosion is calculated in the molecular frame, then the molecular orientation is investigated in the laser electric frame and finally the resulting momenta are expressed in the laboratory frame represented by the axis of the spectrometer. For the sake of simplicity, only the case of the linear triatomic molecule CO_2 in a linearly polarised laser field is presented here. In the molecular frame, the initial positions are $\{R_i\} = \{R_1(\text{C–O}), R_2(\text{O–C}), \gamma = \gamma(\widehat{\text{OCO}})\}$. In order to simplify the calculations, only the symmetric stretching is considered: $R_1(\text{C–O}) = R_2(\text{O–C}) = R$. This approximation is allowed because the symmetric stretching mode with frequency $\nu_1 = 1388.17$ cm^{-1} is excited more easily than is its anti-symmetric counterpart with frequency $\nu_3 = 2349.16$ cm^{-1}. The initial positions $D_{R_i}(R_i)$ can be split into two distributions $D_R(R)$ and $D_\gamma(\gamma)$. The final momenta in the molecular frame are calculated using the classical Newton equations. Since the momenta originate from a three-body fragmentation process, they must lie in one plane. Each momentum is represented by its modulus and its angle with the molecular axis according to $\{P_i\} = \{P_i(R, \gamma), \beta_i(R, \gamma)\}_{i=1,2,3}$.

The orientation of the parent molecule in the laser frame can be represented by an azimuthal angle φ and a polar angle ϑ, where the z axis is defined by the laser field. Whilst the reorientation does not affect the azimuthal distribution $D_\varphi(\varphi)$, it transforms the polar angle distribution $D_\vartheta(\vartheta)$ into one that is strongly peaked along the direction of polarisation of the laser. As mentioned previously, the angle Θ_{pd} between the laser and spectrometer directions is varied in the experiments in order to measure the various components of momenta. The corresponding frame transformation yields the momentum P_{d} for each ion along the axis of the spectrometer:

$$P_{\text{d},i} = P_i(R, \gamma)\{\cos(\Theta_{\text{pd}})\cos[\vartheta + \beta_i(R, \gamma)]$$
$$- \sin(\Theta_{\text{pd}})\sin[\vartheta + \beta_i(R, \gamma)]\sin(\varphi)\}. \qquad (3.20)$$

In equation (3.4), the time of flight of ions should be read with the present notation $T_i = T(P_{\text{d},i} = 0) - P_{\text{d},i}/(ZeF_{\text{c}})$. The multi-dimensional ion spectra, the time-of-flight spectra and the covariance maps are calculated according to equation (3.20) with three meaningful distributions, $D_R(R)$, $D_\gamma(\gamma)$ and $D_\vartheta(\vartheta)$, since the initial isotropic distribution $D_\varphi(\varphi)$ remains unchanged during the explosion.

Figure 3.11 displays an experimental as well as a calculated double-correlation spectrum for CO_2, when the axes of the laser field and the spectrometer are parallel. Owing to the simplicity of the fitting model, triangular distributions are introduced in the calculations. Results obtained with the *same set of parameters* for these distributions are compared with correlation spectra recorded at various angles Θ_{pd} between the directions of the laser field and the spectrometer. The initial geometry of the exploding system is found from an overall agreement, since the physics must be the same, regardless of the detection angle Θ_{pd} [25]. Both the one-dimensional spectrum and the correlation peaks are well reproduced by the calculations presented in figure 3.11. The values obtained from the

Table 3.3. *Fitted geometrical constants of the exploding CO_2 molecule assuming a pure Coulomb repulsion between the multi-charged atomic ions. The excitation conditions are $\lambda = 790$ nm and $I = 10^{15}$ W cm^{-2}. The equilibrium length of the C–O bond is $R_e = 1.162$ Å*

Geometrical value	R	$\Delta R/R$	γ (degrees)	$\Delta\gamma$ (degrees)
Zero-field molecule	R_e	± 0.024	180	± 7.5
Exploding molecule	$2R_e$	± 0.250	180	± 40

fitting procedure are reported in table 3.3 and compared with the values obtained from the unperturbed CO_2 molecule. In accordance with the experimental results given in equation (3.15), the internuclear distance at which the explosion occurs is twice the equilibrium internuclear distance in the Coulomb approximation. The geometry for the explosion remains linear on average. However, large-amplitude nuclear motions appear in table 3.3: the values of $\Delta R/R$ and $\Delta\gamma$ greatly exceed the zero-field values. Calculations of the same type were performed with circular polarisation. In this case, the molecule cannot follow the rapidly rotating electric field and the distribution of orientations is isotropic. Again good agreement with the experimental data was obtained. Other linear molecules, such as N_2O, C_2H_2 and the C_3 chain of C_3H_4 exhibit similar behaviour in strong laser fields.

The geometry of the explosion: linear versus bent molecules

Nuclear motions are more difficult to investigate for bent molecules. Nevertheless, the same procedure allows us to determine the geometry of the explosion, as illustrated in figure 3.12. In the case of CO_2, the ion–ion correlation maxima occur for zero-momentum C^+ ions, since the molecule remains linear on average. In the case of SO_2, on the other hand, the correlation islands exhibit an elliptical profile, which is the signature of a bent system.

Using the model developed above and equation (3.20), the angle $\widehat{OSO}$ is found to be $120 \pm 20°$. This experimental value is similar to the angle $\widehat{OSO}$ of the neutral SO_2 ($\gamma(\widehat{OSO}) = 119.5°$) and singly charged ($\gamma(\widehat{OSO}) = 136.5°$) molecules in their electronic ground states. It is important to point out that the $8a_1$ highest occupied orbital is responsible for the bent structures of SO_2 ($8a_1^2$) and SO_2^+ ($8a_1$). The removal of these electrons does not produce a linear explosion, as one would expect from simple arguments, since SO_2^{2+} is expected to be a linear 16-valence-electron system, just like CO_2. This experimental feature shows that the molecular multiple ionisation does not follow a simple sequential pattern.

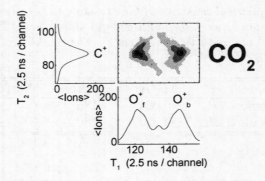

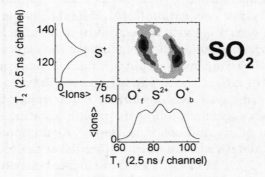

Fig. 3.12. Ion–ion correlation maps recorded at $\lambda = 790$ nm and $F_c = 500$ V cm^{-1} for the C^+/O^+ correlation locations corresponding to the $O^+ + C^+ + O^{Z+}$ ($Z = 1, 2$) fragmentation channels of CO_2 and for the S^+/O^+ correlation locations corresponding to the $O^+ + S^+ + O^{Z+}$ ($Z = 1, 2$) fragmentation channels of SO_2.

3.6 Conclusion and future

In this chapter, we have shown that a simple experimental set-up in association with statistical correlation techniques can resolve many aspects of the fragmentation dynamics of poly-atomic molecules. In particular, the covariance-mapping method introduced by the Reading group is necessary for identifying the multiple-fragmentation channels and overcoming the fact that more than one fragmentation event can occur in strong laser fields. Unfortunately, there is not enough room in this chapter to give an overview of all the techniques developed by other workers. Let us mention the technique of mass-resolved momentum imaging, which appeared recently as a very promising method for determining the momentum distribution of mass-selected fragment ions [29] and gives the possibility of investigating the explosion of vibrationally excited molecules [30]. In addition to the theoretical works presented in the introduction, the calculation of the molecular electron density in strong laser fields that is based on a quantum-chemistry approach might become an essential approach for quantitative predictions [31].

For future experiments, the main efforts will be directed towards increasing the resolution of the experimental measurements. Position-sensitive detectors should be advantageous for use in the simultaneous determination of all the components of the momenta of the ejected fragments. However, it will be easier to work with less than one multiple-fragmentation event per laser shot. In this respect, the advent of very short laser pulses [17] is of crucial importance since it will allow us to reduce the probability of ionisation per laser shot, whilst keeping the same peak laser intensity. In addition, the obtained pulse durations of around 5 fs are smaller than the vibrational times of most molecules. It will thus be possible to study the Coulomb explosion of frozen molecules. In this case, laser excitation could become a small-scale experimental diagnostic for studies of molecular structure, complementary to beam–foil, accelerator-based experiments.

Acknowledgements

Most of the results presented in this chapter were obtained using the Saclay femtosecond-laser facilities. The author is indebted to P. D'Oliveira, O. Gobert, P. Meynadier, M. Perdrix (CEA/SPAM) and G. Vigneron (CEA/SCM) for the successful operation of the LUCA and SOFOCKLE laser systems, as well as to M. Bougeard and E. Caprin for their skilled technical assistance. Several undergraduate students, W. Brochard, J.-J. Esland, L. Ainser and M. Forest, and one graduate student, Ph. Hering, were involved in some of the experiments. Finally the author acknowledges the remarkable editorial work of J. H. Posthumus, due to his deep understanding of molecules and clusters in intense laser fields.

References

[1] G. Herzberg, *Molecular Spectra and Molecular Structure. I. Spectra of Diatomic Molecules*, Van Nostrand Reinhold Company, New York, 1950.

[2] G. Herzberg, *Molecular Spectra and Molecular Structure. III. Electronic Spectra and Electronic Structure of Polyatomic Molecules*, Van Nostrand Reinhold Company, New York, 1966.

[3] Z. Vager, R. Naaman and E. P. Kanter, *Science* **24** 426 (1989).

[4] U. Werner, K. Beckford, J. Becker and H. O. Lutz, *Phys. Rev. Lett.* **74** 1962 (1995).

[5] A. Remscheid, B. A. Huber, M. Pykavyj, V. Staemmler and K. Wiesemann, *J. Phys. B* **29** 515 (1996).

[6] J. H. Posthumus, L. J. Frasinski, A. J. Giles and K. Codling, *J. Phys. B* **28** L349 (1995).

[7] T. Zuo and A. D. Bandrauk, *Phys. Rev. A* **52** R2511 (1995).

[8] S. Chelkowski, T. Zuo, O. Atabek and A. D. Bandrauk, *Phys. Rev. A* **52** 2977 (1995).

[9] T. Seideman, M. Yu. Ivanov and P. B. Corkum, *Phys. Rev. Lett.* **75** 2819 (1995).

[10] M. Brewczyk, K. Rzazewski and C. W. Clark, *Phys. Rev. Lett.* **78** 191 (1997).

[11] H. Sakai, H. Stapelfeldt, E. Constant, M. Yu. Ivanov, D. R. Matusek, J. S. Wright and P. B. Corkum, *Phys. Rev. Lett.* **81** 2217 (1998).

[12] W. C. Wiley and I. H. McLaren, *Rev. Sci. Instrum.* **26** 1150 (1955).

[13] L. J. Frasinski, K. Codling and P. A. Hatherly, *Phys. Lett. A* **142** 499 (1989).

[14] L. J. Frasinski, M. Stankiewitz, K. J. Randall, P. A. Hatherly and K. Codling, *J. Phys. B* **19** L819 (1986).

[15] J. H. D. Eland, F. S. Wort and R. N. Royds, *J. Electron. Spectrosc. Relat. Phenomena* **41** 297 (1986).

[16] M. Simon, T. LeBrun, P. Morin, M. Lavollée and J. L. Maréchal, *Nucl. Instrum. Methods Phys. Res. B* **62** 167 (1991).

[17] M. Nisoli, S. De Silvestri, O. Svelto, R. Szipöcs, K. Ferencz, Ch. Spielmann, S. Sartania and F. Krausz, *Opt. Lett.* **22** 522 (1997).

[18] C. Cornaggia and Ph. Hering, *J. Phys. B* **31** L503 (1998).

[19] R. Thissen, J. Delwiche, J.-M. Robbe, D. Dufflot, J.-P. Flament and J. H. D. Eland, *J. Chem. Phys.* **99** 6590 (1993).

[20] S. D. Price, J. H. D. Eland, P. G. Fournier, J. Fournier and P. Millié, *J. Chem. Phys.* **88** 1511 (1988).

[21] C. Cornaggia, F. Salin and C. Le Blanc, *J. Phys. B* **29** L749 (1996).

[22] C. Cornaggia, J. Lavancier, D. Normand, J. Morellec, P. Agostini, J.-P. Chambaret and A. Antonetti, *Phys. Rev. A* **44** 4499 (1991).

[23] N. Levasseur and P. Millié, *J. Chem. Phys.* **92** 2974 (1990).

[24] C. Cornaggia, *Phys. Rev. A* **52** R4328 (1995).

[25] C. Cornaggia, *Phys. Rev. A* **54** R2555 (1996).

[26] J. H. Posthumus, J. Plumridge, M. K. Thomas, K. Codling, L. J. Frasinski, A. J. Langley and P. F. Taday, *J. Phys. B* **31** L553 (1998).

[27] J. H. Posthumus, J. Plumridge, L. J. Frasinski, K. Codling, A. J. Langley and P. F. Taday, *J. Phys. B* **31** L985 (1998).

[28] Ph. Hering and C.·Cornaggia, *Phys. Rev. A* **59** 2836 (1999).

[29] A. Hishikawa, A. Iwamae, K. Hoshina, M. Kono and K. Yamanouchi, *Chem. Phys. Lett.* **282** 283 (1998).

[30] J. H. Sanderson, R. V. Thomas, W. A. Bryan, W. R. Newel, I. D. Williams, A. J. Langley and P. F. Taday, *J. Phys. B* **31** L59 (1998).

[31] K. Vijayalakshmi, V. R. Bhardwaj, C. P. Safvan and D. Mathur, *J. Phys. B* **30** L339 (1997).

4

Coherent control in intense laser fields

Eric Charron

Laboratoire de Photophysique Moléculaire
Bâtiment 213, Université Paris XI
91405 Orsay Cedex, France
Email: Eric.Charron@ppm.u-psud.fr

Brian Sheehy

National Synchrotron Light Source
Brookhaven National Laboratory
Upton, New York 11973, USA
Email: Sheehy@bnl.gov

4.1 Introduction

The last decade has seen impressive progress in the understanding of the elementary mechanisms of chemical reactions. The advent of femtosecond (fs) laser sources has allowed the possibility of analysing ultra-fast chemical reactions using lasers [1]. In *pump–probe* schemes, a first ultra-short pulse initiates a reaction that is monitored by a subsequent probe laser pulse. This method is extremely useful for recording the formation and the evolution of transition states in real time. With the understanding of these fundamental mechanisms, there has emerged a renewal of interest in the control of photo-induced processes using tailored laser pulses [2].

The ability to control chemical reactions through a variety of means (catalysis, use of various substrates, variation of temperature, pressure, concentration of the initial reactants, etc.) has major beneficial consequences for chemical synthesis and other industrial manufacturing processes. Not only is it normally useful to speed up a reaction but also, in a case with more than one possible outcome, it can be extremely useful to influence the branching ratios of the various fractions. This can result in a far greater yield of the desired product, with the elimination of by-products. Classical methods of chemistry sometimes fail to produce a specific product from a given set of reactants. This is the case, for instance, when the reaction is under thermodynamic control and one wants to produce a metastable species. There is hope that a new laser-control scenario could be used to modify the natural path and outcome of such a reaction.

The traditional approach of just using wavelength variations of a CW laser in the weak-field domain has proven to be rather limited for achieving

114

these goals. More recently, different approaches using the unique characteristics of new laser sources have been proposed in order to surmount these difficulties.

The *time-domain approach*, initially proposed by D. J. Tannor and S. A. Rice [3], uses a succession of tailored and short (usually in the femtosecond domain) laser pulses. Using this scheme, one can in principle create a well-defined wavepacket whose subsequent driven evolution will lead to a specific state of the molecule or to a specific dissociation or ionisation channel, thus creating the desired product. This method has proven to be efficient in a certain number of simple situations. However, when the molecular dynamics is rather complex, producing the required sequence of tailored pulses can be a difficult experimental challenge.

On the theoretical side, one is then led to apply optimal-control theory [4] for the determination of the best temporal and spectral profiles to maximise the yield of the desired product. This approach has been investigated thoroughly by several groups [4], but a detailed discussion is beyond the scope of the present chapter. Feedback-controlled [5] experiments [6] have recently given very promising results on rather complex systems with this method.

Various experimental [1, 7–9] and theoretical [1, 3, 10] demonstrations of this time-domain method have been implemented, usually with weak fields. When the potential-energy surfaces of the reactant are well known, it is easier to design intuitively a sequence of pulses to induce the desired dynamics. This is the case, for example, in the experiment of T. Baumert and G. Gerber [7], in which two short pulses are able to control efficiently the competition between the ionisation and dissociative-ionisation channels of the Na_2 molecule.

As noted by P. Brumer and M. Shapiro [11], in the weak-field domain it is impossible to induce any active control just by varying the duration of a single pulse. With intense fields, this limitation is suppressed due to the non-perturbative character of the interaction and the timescale of excitation becomes relevant for the control of the photofragmentation dynamics. This effect has been examined by M. Machholm and A. Suzor-Weiner [10], who used it to control the dissociation channel chosen by an excited Na_2^+ nuclear wavepacket, to form either $Na(3s)+Na^+$ or $Na(3p)+Na^+$, depending on how long the pulse is acting.

A so-called *incoherent interference control* scheme has also been proposed by Z. Chen *et al.* [12], using the phenomenon of laser-induced continuum structures. In this scheme, one uses a strong CW laser to impose a given structure on the continuum by dressing it with an initially unpopulated bound state. Another pulse (not necessarily strong) is then used to promote a populated bound state to this continuum. An experimental study by A. Schnitman *et al.* [13] clearly showed that, although there is no phase relation between the two light sources, quantum interferences among reaction pathways are responsible for the control of the competition between the Na(3s) + Na(3d) and Na(3s) + Na(3p) channels in the photodissociation of Na_2. Varying the laser wavelengths induces an important change in the corresponding branching ratios.

In this chapter we will concentrate on the *frequency-domain approach* proposed initially by P. Brumer and M. Shapiro [14]. This method, called the *coherent control* method, also uses the principle of quantum interferences among reaction pathways. In section 4.2, we discuss its basic principles and its various implementations in the weak-field domain. Section 4.3 shows how it extends to the strong-field limit. In section 4.4 we summarise the theory of wavepackets in strong fields and in section 4.5 we present some theoretical examples of strong-field coherent control of photodissociation of the simple species H_2^+ and HD^+. These results are compared with available experimental data in section 4.6. Finally, section 4.7 presents some interesting perspectives for future directions.

4.2 Coherent control in the weak-field limit

Introduction

The *coherent control* method [14] is based on the interference among reaction paths due to the *coherent superposition* of laser radiation at the fundamental frequency and one of the harmonics, as shown in figure 4.1 for the case of the third harmonic. Two paths for the excitation of the initial state **M** of the molecule to form the transition state $M^{\neq}$ are available, leading finally to the different reaction channels $\{\alpha, \beta, \ldots\}$. In the simplest case of the third harmonic, the first reaction path involves the absorption of three photons $\hbar\omega_1$ of the fundamental while the second path is characterised by the absorption of a single photon $\hbar\omega_3$ of the harmonic. The probability of ending up in each channel can be formally written with the help of the probability amplitudes associated with each absorption path. The probability of exiting in channel α can, for instance, be written as

$$P_\alpha = \left| \sum_{\text{paths}} A_{\alpha,i} \right|^2 = |A_{\alpha,3\omega_1} + A_{\alpha,\omega_3}|^2. \qquad (4.1)$$

The total probability P_α is thus sensitive to the term $\Re\left(A_{\alpha,3\omega_1}^* A_{\alpha,\omega_3} \right)$ describing interference between the two excitation paths. The sign and the amplitude of this cross term are strongly dependent on the laser parameters which are controllable in the laboratory.

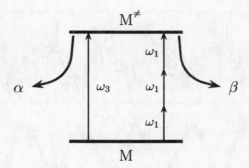

Fig. 4.1. The general principle of coherent control: superposition of a fundamental laser frequency ω_1 and its third harmonic $\omega_3 = 3\,\omega_1$.

In its classical formulation, the corresponding linearly polarised laser electric field is represented by

$$\boldsymbol{E}(t) = E(t)\hat{\mathbf{e}} = f(t)[E_1\cos(\omega_1 t) + E_3\cos(\omega_3 t + \varphi)]\hat{\mathbf{e}}, \qquad (4.2)$$

where $f(t)$ refers to the temporal envelope of the pulse, $\hat{\mathbf{e}}$ is the unitary polarisation vector and φ is the phase shift (fixed in the experiment) between the two harmonics. In the limit of weak fields it can be shown, with the help of the third-order perturbation theory [15], that the branching ratio P_α/P_β depends only on the following control parameters: the phase lag φ and the ratio of the amplitudes of the two electric fields E_3/E_1^3. By changing these parameters one modifies the interference term $\Re\!\left(A^*_{\alpha,3\omega_1} A_{\alpha,\omega_3}\right)$ and determines its constructive or destructive nature. One can thus significantly increase or decrease the rate of formation of a given reaction product.

Coherent control of the reaction probabilities

Although the manipulation of the electronic and nuclear properties of molecules using the coherent control scheme presented above is still at the stage of fundamental research, it has been applied successfully in many theoretical and experimental studies (see [2] and references therein).

The aim of the first experimental studies was simply to test the validity of this new method of control by demonstrating the variation of the excitation probabilities of simple atomic and molecular systems with the phase shift φ between the two harmonics, see equation (4.2). These experiments have revealed characteristic oscillations in fluorescence signals [16], in the ion or electron signals obtained by multiphoton ionisation of various atoms [17, 18], diatomic molecules [19] and polyatomic species [20, 21] and in rates of dissociation of molecules [22].

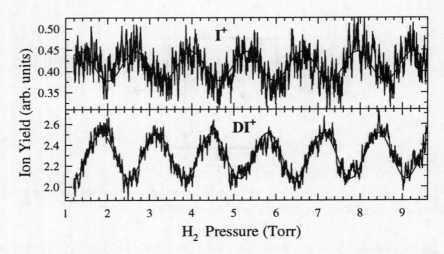

Fig. 4.2. I^+ and DI^+ signals as a function of the pressure of H_2 in a phase-tuning cell, i.e. as a function of the phase lag φ between the fundamental radiation $\lambda_1 = 353.69$ nm and its third harmonic $\lambda_3 = 117.90$ nm. Two I^+ or DI^+ maxima are separated by $\Delta\varphi = 2\pi$ (see section 4.3). Reproduced from [23] with the permission of the authors.

In these experiments, it is crucial to maintain the coherence between the two wavelengths used. The harmonic is thus usually obtained by tripling the fundamental radiation and one varies the phase shift φ between the harmonic and the fundamental by changing the optical lengths of their paths to the experiment.

Until now, no systematic investigation of the control parameters has been performed experimentally and it seems difficult to predict *a priori* the *optimal* conditions of control. However, a 75% modulation of the probabilities of ionisation of CH_3I, achieved by varying only the phase shift φ, has been reported [21]. V. Blanchet *et al.* [8] recently extended this approach to coherent control by using a sequence of two coherent laser pulses of the same wavelength.

A very convincing experimental implementation of the coherent control scheme with various fragmentation channels in competition has been realised by L. Zhu *et al.* [23] for the molecules HI and DI. This experiment demonstrated that interferences induced by the presence of various excitation pathways can modify significantly the branching ratio of two channels that are clearly chemically distinguishable. The molecular system is excited simultaneously by one vacuum-ultraviolet photon and by three ultraviolet photons above the first ionisation threshold. By varying the pressure of the H_2 gas located in a phase-tuning cell, the phase shift φ between the two colours is modulated over a few 2π periods. De-

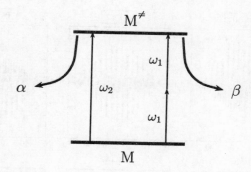

Fig. 4.3. Coherent control using harmonics of opposite parities: superposition upon a fundamental laser frequency ω_1 of its second harmonic $\omega_2 = 2\omega_1$.

pending on the value of φ, DI molecules decay preferentially either by auto-ionisation to form DI^+ or by predissociation to produce I^+ after a second ionisation step. Figure 4.2 clearly shows that the DI^+ and I^+ signals are out of phase, so the competition between the dissociative and ionisation channels can be controlled by changing φ.

Coherent control of the angular distributions

Modulating the probabilities P_α in various fragmentation channels requires the same symmetry for the state $M^{\neq}$ of figure 4.1, whatever the excitation path. If this criterion is not respected, rules specifying orthogonality between states of different symmetries lead to an incoherent sum of the probabilities associated with each excitation channel in the calculation of the integrated probability of exiting in a given channel α:

$$P_\alpha = \sum_{\text{paths}} |A_{\alpha,i}|^2 = |A_{\alpha,3\omega_1}|^2 + |A_{\alpha,\omega_3}|^2. \tag{4.3}$$

In this type of incoherent sum, the cross terms describing the interference among reaction paths seen in equation (4.1) disappear. The probabilities P_α no longer depend on the phase shift φ between the two colours, so a very efficient control parameter is lost. This is why the initial proposal of Brumer and Shapiro was based on a superposition of two harmonics of the same parity (e.g. ω_1 and ω_3).

However, harmonics of opposite parities (ω_1 and $\omega_2 = 2\omega_1$ for instance, see figure 4.3) may also induce some important interference effects if one is interested in differential cross sections such as the angular distributions of the reaction products. The most striking feature is the asymmetry observed in the angular dependence of the reaction probabilities.

As can be seen in figure 4.4, the total two-colour electric field

$$\boldsymbol{E}(t) = E(t)\hat{\mathbf{e}} = f(t)[E_1 \cos(\omega_1 t) + E_2 \cos(\omega_2 t + \varphi)]\hat{\mathbf{e}} \tag{4.4}$$

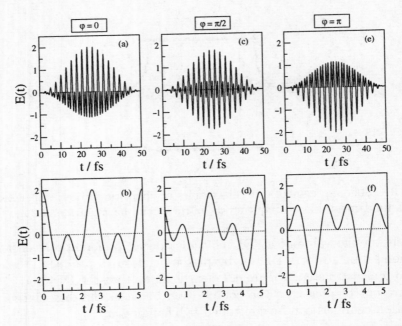

Fig. 4.4. The two-colour electric field $E(t)$ of equation (4.4), with a fundamental wavelength of 800 nm and its second harmonic. In panels (a), (c) and (e), the pulse shape $f(t)$ is $\sin^2[\pi t/(2T_{\rm p})]$ with a pulse duration $T_{\rm p}$ of 25 fs. In panels (b), (d) and (f), $f(t)$ is constant. The phase shift between the fundamental and the second harmonic is $\varphi = 0$ in (a) and (b), $\varphi = \pi/2$ in (c) and (d) and $\varphi = \pi$ in (e) and (f).

exhibits a clear asymmetry if $\varphi \neq \pi/2$ (modulo π). This effect is directly reflected in the angular distributions, showing a clear difference between the probabilities of ejection of the fragments in the direction $\hat{\boldsymbol{k}} \equiv (\theta_k, \phi_k)$ and in the opposite direction $-\hat{\boldsymbol{k}} \equiv (\pi - \theta_k, \pi + \phi_k)$, where $\hat{\boldsymbol{k}}$ denotes the unitary wave vector of the fragments. This phenomenon has been predicted [24] and measured [25] for control of the directionality of the photocurrent in semiconductors. It has then been applied to various situations of gas-phase photofragmentation, either theoretically [26, 27] or experimentally [18, 28, 29], and has been proposed as an alternative solution for the orientation of heteronuclear molecules [30].

4.3 Coherent control in the strong-field limit

In intense laser fields the coherent control method has been used to achieve some degree of control over such processes as photo-ionisation [31], high-harmonic generation [32, 33], above-threshold ionisation [18, 33, 34] and photodissociation [29, 35–40]. In the present section we describe how the

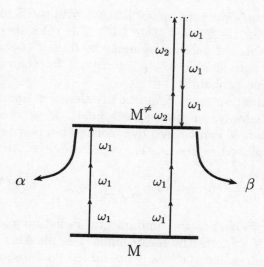

Fig. 4.5. A schematic example of two excitation mechanisms in competition using the coherent control method with harmonics of opposite parities in intense fields.

coherent control scheme extends to the strong-field limit. The reader who is more concerned with specific theoretical and experimental examples of intense-field coherent control of photodissociation may wish to proceed directly to sections 4.5 and 4.6.

In the intense-field regime the lowest-order perturbation theory becomes irrelevant: phenomena of much higher order can sometimes not only compete with the lowest-order processes but even dominate the whole dynamics. As a consequence it is no longer possible to consider the simple picture of figure 4.1 as the unique mechanism of molecular excitation. In the case of a superposition upon the fundamental frequency ω_1 of its third harmonic ω_3, the branching ratio P_α/P_β now depends not on the phase shift φ and on a ratio of electric-field amplitudes E_3/E_1^3 but rather on these three parameters independently. One thus gains an additional control parameter at the cost of losing the simplicity of weak-field interactions.

Additionally the restriction of using only frequencies of the same parity to achieve coherent control of a total probability of fragmentation is lifted. This effect can be understood by looking, for example, at the two excitation schemes shown in figure 4.5 for the case of the fundamental with its second harmonic. The first path is simply a three-photon absorption of the fundamental frequency, while the second path involves the absorption of two photons of frequency ω_1 and two photons of frequency ω_2, followed by stimulated emission of three more photons of frequency ω_1. In both cases odd numbers of photons are exchanged with the field, leading to the same symmetry for the excited state $\mathbf{M}^{\neq}$. The rules specifying orthogonality between states of different symmetries do not apply and one is left with a coherent sum similar to equation (4.1) for the calculation of an integrated reaction probability.

We will now analyse these complex interferences among many different excitation pathways in intense laser fields. If we consider a linearly polarised electric field constituted by a coherent superposition of two frequencies, its quantised representation is given by the operator

$$\tilde{\mathcal{E}} = \frac{i}{2\sqrt{\bar{N}_f}}\left(\tilde{a}_f - \tilde{a}_f^{\dagger}\right)\hat{\mathbf{e}} + \frac{i}{2\sqrt{\bar{N}_h}}\left(\tilde{a}_h\,e^{i\varphi} - \tilde{a}_h^{\dagger}\,e^{-i\varphi}\right)\hat{\mathbf{e}}, \qquad (4.5)$$

where the index f refers to the fundamental radiation and the index h to the harmonic. $\bar{N}$ denotes the mean number of photons in the field and $\tilde{a}$ ($\tilde{a}^{\dagger}$) is the destruction (creation) operator. As in equation (4.2), the phase shift φ is associated with the operators describing the harmonic electric field. Since absorption and stimulated emission are described by the operators $\tilde{a}$ and $\tilde{a}^{\dagger}$, it is clear from equation (4.5) that the wavefunction describing the molecular system is multiplied by the phase factor $e^{i\varphi}$ during the absorption of a photon of the harmonic. On the other hand, when the molecule emits one photon of the harmonic, its wavefunction is multiplied by $e^{-i\varphi}$ and is thus dephased by the opposite quantity.

For simplicity, let us first consider that we are in a situation such that the two excitation mechanisms shown in figure 4.5 are dominating our dynamics. If we look only at the harmonic, these two pathways simply differ by the absorption of two photons of frequency ω_2. The probability amplitude associated with the second excitation pathway is thus dephased by $e^{i\,2\varphi}$ relative to the amplitude associated with the first pathway. As a result, the probability of exiting in channel α can be written as the following sum of two independent probabilities associated with each excitation pathway plus an interference term showing explicitly the phase dependence:

$$P_{\alpha} = \left|\sum_{\text{paths}} A_{\alpha,i}\right|^2 = |A_{\alpha,1}|^2 + |A_{\alpha,2}|^2 + 2\,\Re\mathrm{e}\left(A_{\alpha,1}^* A_{\alpha,2} e^{i\,2\varphi}\right). \qquad (4.6)$$

The probability P_{α} thus oscillates as $\cos(2\varphi)$, with a periodicity of π.

For the general case of an infinite number of open excitation paths, this result can easily be generalised [39] if one takes into account the fact that only pathways ending up at the same energy with the same symmetry may interfere. In a commensurate-frequency case in which the ratio $x = \omega_h/\omega_f$ of the harmonic and fundamental angular frequencies is

Table 4.1. *Phase periods of P_α for the first five values of q_x (see the text)*

$x = \omega_{\mathrm{h}}/\omega_{\mathrm{f}}$	q_x	Phase period of P_α (radians)
1, 3, 5, 7 ...	1	2π
2, 4, 6, 8 ...	2	π
5/3, 7/3, 11/3 ...	3	$2\pi/3$
3/2, 5/2, 7/2 ...	4	$\pi/2$
7/5, 9/5, 11/5 ...	5	$2\pi/5$

rational, the dependence on φ of the reaction probability can be written as

$$P_\alpha = \sum_{n=0}^{\infty} p_n \cos(n q_x \varphi), \qquad (4.7)$$

where q_x denotes the first non-zero positive integer such that $q_x(x-1)/2$ is an integer. In this expression p_0 represents the sum of the probabilities due to each excitation pathway and p_n $(n \geq 1)$ is the weight of the term describing interference among pathways that differ by the exchange (emission or absorption) of $n q_x$ photons of the harmonic. The period of oscillation of P_α with φ is then given by $2\pi/q_x$. For the first five values of q_x the phase periods of P_α are indicated in table 4.1, together with the associated ratio of frequencies $\omega_{\mathrm{h}}/\omega_{\mathrm{f}}$. For even harmonics, like ω_2 for instance, the sum (4.7) is limited to even numbers $n q_x$ of harmonic photons (because $q_x = 2$) and the phase period of P_α is π. On the other hand, no restriction is imposed on equation (4.7) for odd harmonics, such as ω_3 (because in this case $q_x = 1$) and the period of P_α is 2π.

The phase period of the reaction probability P_α is also plotted in figure 4.6 as a function of $\omega_{\mathrm{h}}/\omega_{\mathrm{f}}$. One may notice in this picture the five examples given in table 4.1. As demonstrated by the scaled inset of this figure, this phase period has a fractal structure, with a calculated Hausdorff–Besicovich dimension [41] of 1.29.

In theory, only commensurate frequencies may induce some interferences because incommensurate frequencies never reach the same final energy. In practice, one always finds commensurate frequencies within the bandwidths of the fundamental and harmonic laser pulses. Frequencies that are *a priori* incommensurate may thus induce some interference effects in the reaction probabilities. As a consequence, the picture drawn in figure 4.6 could be measured experimentally only with a certain degree of convolution, defined by the finite pulse durations.

Using similar arguments, one can show that the probability of photo-

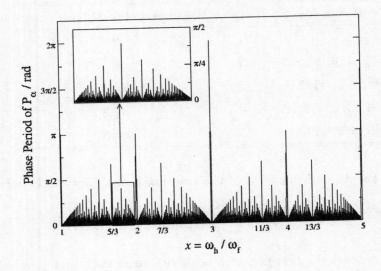

Fig. 4.6.   Phase period of P_α as a function of ω_h/ω_f.

fragmentation in a well-defined direction $\hat{\boldsymbol{k}} \equiv (\theta_k, \phi_k)$ can also be written as the sum given in equation (4.7), where q_x now denotes the first non-zero positive integer such that $[q_x(x-1)-1]/2$ is an integer. The corresponding period of oscillation of $P_\alpha(\hat{\boldsymbol{k}})$ with φ is $2\pi/q_x$. For the first five values of q_x the phase periods of the differential probability $P_\alpha(\hat{\boldsymbol{k}})$ are indicated in table 4.2, with the associated ratio ω_h/ω_f.

A picture of the oscillation period of $P_\alpha(\hat{\boldsymbol{k}})$ as a function of ω_h/ω_f would show a fractal structure very similar to the one obtained in figure 4.6, with an equal dimension. In addition, exchanging φ for $\varphi + \pi/q_x$ leads to an

Table 4.2.   *Phase periods of $P_\alpha(\hat{\boldsymbol{k}})$ for the first five values of q_x (see the text)*

$x = \omega_h/\omega_f$	q_x	Phase period of $P_\alpha(\hat{\boldsymbol{k}})$ (radians)
2, 4, 6, 8 …	1	2π
3/2, 5/2, 7/2 …	2	π
4/3, 8/3, 10/3 …	3	$2\pi/3$
5/4, 7/4, 9/4 …	4	$\pi/2$
6/5, 8/5, 12/5 …	5	$2\pi/5$

inversion of the angular distributions of fragments, i.e. $P_\alpha(\hat{\boldsymbol{k}})$ becomes $P_\alpha(-\hat{\boldsymbol{k}})$ and *vice versa*.

In conclusion it appears that the coherent control scenario, introduced by P. Brumer and M. Shapiro [14] in the case of a weak interaction between the molecular target and the laser pulse, can be extended to the strong-field regime at the evident cost of losing its simplicity. On the other hand, strong fields open higher-order excitation pathways that add flexibility helping to control branching ratios efficiently, with very high total yields. The next section introduces the time-dependent theory that we will use in section 4.5 to treat some examples of strong field coherent control of photodissociation, in the case of the hydrogen molecular ion.

4.4 A time-dependent theory of strong-field photodissociation

When the molecular excitation is performed with a pulse of finite duration, the electro-nuclear Hamiltonian $\mathcal{H}(\boldsymbol{r}, \boldsymbol{R}, t)$ is explicitly time-dependent, so we must work with the time-dependent Schrödinger equation

$$i\hbar \frac{\partial \Psi(\boldsymbol{r}, \boldsymbol{R}, t)}{\partial t} = \mathcal{H}(\boldsymbol{r}, \boldsymbol{R}, t)\Psi(\boldsymbol{r}, \boldsymbol{R}, t), \qquad (4.8)$$

where we denote here the complete sets of three-dimensional coordinates of all electrons and nuclei by $\boldsymbol{r}$ and $\boldsymbol{R}$, respectively, and $\Psi(\boldsymbol{r}, \boldsymbol{R}, t)$ is the time-dependent electro-nuclear wavefunction.

While the interaction between an external electromagnetic field and an atom or a molecule is usually treated perturbatively for weak laser pulses, non-perturbative treatments are necessary when the interaction energy becomes comparable to the target's binding energy. In this context, theoretical descriptions including absorption and stimulated emission as well as the numerous electronic and nuclear continua involved are extremely difficult, so numerical grid techniques are usually chosen. Additionally, solving the time-dependent Schrödinger equation (4.8) for the electro-nuclear wavefunction $\Psi(\boldsymbol{r}, \boldsymbol{R}, t)$ gives a direct picture of the excitation dynamics.

The time-dependent electro-nuclear Hamiltonian can be developed as

$$\mathcal{H}(\boldsymbol{r}, \boldsymbol{R}, t) = T_e(\boldsymbol{r}) + T_N(\boldsymbol{R}) + V_{\text{Coulomb}}(\boldsymbol{r}, \boldsymbol{R}) + V_{\text{int}}(\boldsymbol{r}, \boldsymbol{R}, t), \qquad (4.9)$$

where $T_e(\boldsymbol{r})$ and $T_N(\boldsymbol{R})$ denote the electronic and nuclear kinetic-energy operators and $V_{\text{Coulomb}}(\boldsymbol{r}, \boldsymbol{R})$ is the electron–electron, electron–nuclei and nuclei–nuclei Coulomb interaction. The energy of interaction between the field and the molecular system $V_{\text{int}}(\boldsymbol{r}, \boldsymbol{R}, t)$ can be represented either in the length (i.e Lorentz) or in the velocity (i.e. Coulomb) gauge. For an exact calculation this choice would have no consequence, but practically, as soon as an approximation is introduced, the choice of gauge can change the results of the numerical simulations considerably. This is especially true if a truncated basis representation is used for the electro-nuclear wavepacket $\Psi(\boldsymbol{r}, \boldsymbol{R}, t)$. Throughout this chapter we will make use of the length gauge, although it is not easy to know *a priori* which gauge gives the best representation of the dynamics. For the special case of the molecular hydrogen ions H_2^+ and HD^+ (see section 4.5), it has been proven that this choice is accurate even when only two electronic states are included, as long as the ionisation of the molecule is negligible [38].

In the electric-dipole approximation, the molecular system is coupled to the time-dependent electric field $\boldsymbol{E}(t)$ via the electric dipole $\boldsymbol{\mu}(\boldsymbol{r}, \boldsymbol{R})$:

$$V_{\text{int}}(\boldsymbol{r}, \boldsymbol{R}, t) = -\boldsymbol{\mu}(\boldsymbol{r}, \boldsymbol{R}) \cdot \boldsymbol{E}(t). \tag{4.10}$$

Using the adiabatic approach, we now define a complete set of electronic states represented by the eigenfunctions $\varphi_n(\boldsymbol{r}|\boldsymbol{R})$ of the electronic Hamiltonian $\mathcal{H}_e(\boldsymbol{r}, \boldsymbol{R}) = T_e(\boldsymbol{r}) + V_{\text{Coulomb}}(\boldsymbol{r}, \boldsymbol{R})$ for a fixed nuclear configuration $\boldsymbol{R}$:

$$\mathcal{H}_e(\boldsymbol{r}, \boldsymbol{R})\varphi_n(\boldsymbol{r}|\boldsymbol{R}) = V_n(\boldsymbol{R})\varphi_n(\boldsymbol{r}|\boldsymbol{R}). \tag{4.11}$$

The exact Born–Oppenheimer expansion

$$\Psi(\boldsymbol{r}, \boldsymbol{R}, t) = \sum_n \chi_n(\boldsymbol{R}, t)\varphi_n(\boldsymbol{r}|\boldsymbol{R}) \tag{4.12}$$

then associates a unique nuclear wavepacket $\chi_n(\boldsymbol{R}, t)$ with each electronic potential-energy surface $V_n(\boldsymbol{R})$. In accordance with the usual assumption of a weak coupling between the electronic and nuclear degrees of freedom, we now apply the Born–Oppenheimer approximation, neglecting the variation of the adiabatic electronic wavefunctions $\varphi_n(\boldsymbol{r}|\boldsymbol{R})$ with $\boldsymbol{R}$. This approximation allows us to separate clearly the nuclear and electronic dynamics. Inserting the expansion (4.12) into the time-dependent Schrödinger equation (4.8) and projecting the resulting expression onto the adiabatic electronic state $\varphi_m(\boldsymbol{r}|\boldsymbol{R})$ gives the time-dependent equations obeyed by the nuclear wavepackets $\chi_m(\boldsymbol{R}, t)$:

$$i\hbar\frac{\partial\chi_m(\boldsymbol{R}, t)}{\partial t} = (T_N(\boldsymbol{R}) + V_m(\boldsymbol{R}))\chi_m(\boldsymbol{R}, t)$$
$$+ \sum_n \mu_{mn}(\boldsymbol{R})E(t)\chi_n(\boldsymbol{R}, t), \tag{4.13}$$

where $\mu_{mn}(R)$ denotes the transition dipole for the electronic states n and m

$$\mu_{mn}(\boldsymbol{R}) = \langle\varphi_m(\boldsymbol{r}|\boldsymbol{R})| - \boldsymbol{\mu}(\boldsymbol{r}, \boldsymbol{R}) \cdot \hat{\mathbf{e}}|\varphi_n(\boldsymbol{r}|\boldsymbol{R})\rangle_{\boldsymbol{r}}. \tag{4.14}$$

One can then solve the system of time-dependent coupled differential equations (4.13), knowing the initial state of the system at time $t = 0$, with various numerical techniques [42]. The split-operator method [43] is often chosen in intense fields for the simplicity of its implementation, its accuracy and its efficiency, which mainly come from the use of fast Fourier transforms. In this scheme, the initial wavefunction is propagated in time throughout the duration of the pulse $(0 \leq t \leq t_f)$

$$
\begin{pmatrix} \chi_1(\boldsymbol{R}, t_f) \\ \vdots \\ \chi_m(\boldsymbol{R}, t_f) \\ \vdots \end{pmatrix} = U(t_f \leftarrow 0) \begin{pmatrix} \chi_1(\boldsymbol{R}, 0) \\ \vdots \\ \chi_m(\boldsymbol{R}, 0) \\ \vdots \end{pmatrix}, \qquad (4.15)
$$

using the propagator $U(t_f \leftarrow 0)$ which can be decomposed into a product of short-time propagators acting during a small time interval δt:

$$
U(t_f \leftarrow 0) = U(t_f \leftarrow t_f - \delta t) \times \cdots \times U(\delta t \leftarrow 0). \qquad (4.16)
$$

These short-time propagators $U(t \leftarrow t - \delta t)$ are then split as

$$
U(t \leftarrow t - \delta t) \approx \exp\left(-i\frac{T\,\delta t}{2\hbar}\right) \times \exp\left(-i\frac{V\,\delta t}{\hbar}\right) \times \exp\left(-i\frac{T\,\delta t}{2\hbar}\right) \qquad (4.17)
$$

and this separates the potential- and kinetic-energy terms to within an accuracy of the order of $(\delta t)^2$. In this equation, V denotes the potential matrix containing the electronic diagonal potentials and the non-diagonal radiative couplings (see equation (4.13)), while T stands for the diagonal matrix of the nuclear kinetic-energy operator.

Once the propagation in time has been performed, one remains with a set of nuclear wavepackets $\chi_m(\boldsymbol{R}, t_f)$ that are analysed by projection onto the bound or dissociated stationary states of the molecule to get the total and differential probabilities of dissociation (see [35, 44] for a more detailed description).

4.5 Simulations of photodissociation

In this section we examine the effect of a coherent superposition of two harmonics of the same laser (see equation (4.2) for the case of the first and third harmonics) on the probabilities of dissociation of the hydrogen molecular ions H_2^+ and HD^+. The simple electronic structures of these molecules allow a precise comparison between experiment and theory and provide a first basis for the extension of control to more complex systems.

Depending on the relative phase φ between the two colours, the maxima of the oscillations describing the two electric fields will be either shifted or superimposed. If they are superimposed, they have either the same or opposite signs. In the first case, the total amplitude of the electric field is

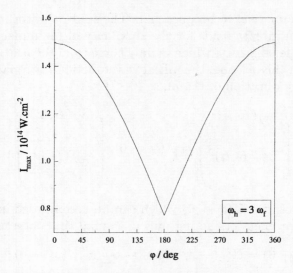

Fig. 4.7. The peak intensity reached during the laser pulse as a function of the phase lag φ between the fundamental radiation $\lambda_f = 780$ nm and the third harmonic $\lambda_h = 260$ nm with the intensities $I_f = 10^{14}$ W cm^{-2} and $I_h = 5 \times 10^{12}$ W cm^{-2}.

greater than that of each harmonic, whereas in the second case the global amplitude is less than the amplitude of each colour. For the third harmonic the first possibility is observed when $\varphi \equiv 0$ (modulo 2π), whereas the second possibility corresponds to the phase shift $\varphi \equiv \pi$ (modulo 2π). This effect is illustrated in figure 4.7, which shows the maximum total intensity reached during the laser pulse for the following laser parameters: $\lambda_f = 780$ nm, $\lambda_h = 260$ nm, $I_f = 10^{14}$ W cm^{-2} and $I_h = 5 \times 10^{12}$ W cm^{-2}.

In the non-linear regime the molecular dynamics is usually governed by the radiative intensity imposed on the system. If this hypothesis is strictly respected, it means that a higher rate of dissociation should be observed for $\varphi = 0$ than for $\varphi = \pi$. Actually, this is not always the case. For instance, figure 4.8 shows the probability of dissociation of H_2^+ as a function of φ for the laser parameters used to draw figure 4.7. The initial vibrational level is $v_0 = 0$ in the left-hand part of the figure and $v_0 = 1$ on the right-hand side. These results have been obtained in a one-dimensional approximation, whereby the molecular ion is assumed to be aligned along the axis of polarisation of the laser throughout the duration of the pulse. This approximation is actually justified by the efficient alignment of the molecule by the intense linearly polarised pulse (see the experimental section 4.6). In this figure, the squares represent the numerical results of the time-dependent simulation while the solid line is given by the sum in equation (4.7), where the parameters p_n have been

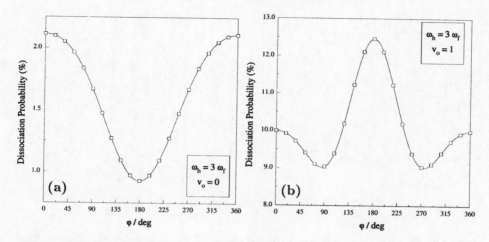

Fig. 4.8. The probability of dissociation of H_2^+ as a function of the phase lag φ between the fundamental radiation $\lambda_f = 780$ nm and the third harmonic $\lambda_h = 260$ nm with the intensities $I_f = 10^{14}$ W cm^{-2} and $I_h = 5 \times 10^{12}$ W cm^{-2}. The pulse width is $T_p = 150$ fs. The initial level is $v_0 = 0$ in (a) and $v_0 = 1$ in (b).

fitted to the numerical points.

From figure 4.8, it is clear that the probability of dissociation from the initial level $v_0 = 0$ varies as the total intensity of figure 4.7. On the other hand, the initial level $v_0 = 1$ has the opposite behaviour. These seemingly contradictory behaviours originate from effects of quantum interference among reaction pathways. One needs to include these coherent phenomena in a model description of the dissociation process; relying on a simple picture of the total electric field might be insufficient.

A better picture can be obtained by constructing a complete set of adiabatic potential curves [27] (see also chapter 2), which describes not only the electronic structure of the molecule but also the interaction of the laser with all of the possible excitation paths. One can demonstrate [35] that the molecular structure is altered differently by the laser for $\varphi = 0$ and for $\varphi = \pi$. Because the $v_0 = 0$ and $v_0 = 1$ levels have different energies, they are affected differently.

In section 4.3, we have shown that a characteristic of the intense-field regime for coherent control is that even a coherent superposition of two harmonics of opposite parities can create some interference effects, which will induce a variation of the total probability of dissociation with the phase shift φ. An example of this phenomenon is shown in figure 4.9. This figure shows the probability of dissociation of H_2^+ as a function of φ for a coherent superposition upon the first harmonic $\lambda_f = 780$ nm of the second harmonic $\lambda_h = 390$ nm in (a) and of the commensurate 9/7th

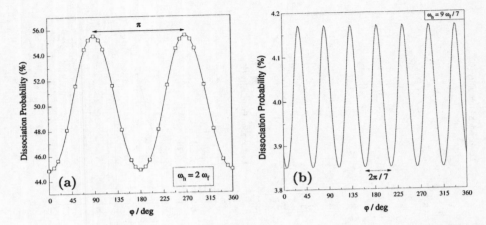

Fig. 4.9. (a) The probability of dissociation of H_2^+ as a function of the phase lag φ between the fundamental radiation $\lambda_f = 780$ nm and the second harmonic $\lambda_h = 390$ nm with the intensities $I_f = 10^{14}$ W cm^{-2} and $I_h = 3 \times 10^{13}$ W cm^{-2}. The pulse width is $T_p = 150$ fs and the initial level $v_0 = 1$. (b) The probability of dissociation of H_2^+ as a function of the phase lag φ between the fundamental radiation $\lambda_f = 780$ nm and its commensurate 9/7th frequency $\lambda_h = 606.7$ nm with the intensities $I_f = 10^{14}$ W cm^{-2} and $I_h = 3 \times 10^{13}$ W cm^{-2}. The pulse width is $T_p = 150$ fs and the initial level $v_0 = 1$.

frequency $\lambda_h = 606.7$ nm in (b). In both cases the pulse lasts 150 fs, the initial state is $v_0 = 1$ and the intensities are $I_f = 10^{14}$ W cm^{-2} and $I_h = 3 \times 10^{13}$ W cm^{-2}. Again, the squares represent the numerical results and the solid line is given by the sum in equation (4.7), where the parameters p_n have been fitted to the numerical points, with the restriction that $q_x = 2$ for the second harmonic and $q_x = 7$ for the 9/7th harmonic (see table 4.1). In this figure, one clearly sees the periods π and $2\pi/7$ predicted in section 4.3 for these two cases.

As shown in figure 4.4, a superposition upon a fundamental wavelength of its second harmonic yields an asymmetric electric field. This obviously has a strong impact on the angular distributions of the fragments. As seen in section 4.3 (see table 4.2), this can also happen for some specific commensurate frequencies.

This phenomenon is illustrated by figure 4.10, which shows the probability of dissociation of HD^+ in the two lowest channels $H^+ + D(1s)$ (solid line) and $H(1s) + D^+$ (dashed line) in the forward (parts (a)) and backward (parts (b)) directions as a function of the phase shift between the fundamental radiation and its second harmonic. Here, forward stands for $\theta_k = 0$, i.e. the direction of the ion detector (see the experimental section 4.6), while backward is the opposite direction, $\theta_k = \pi$.

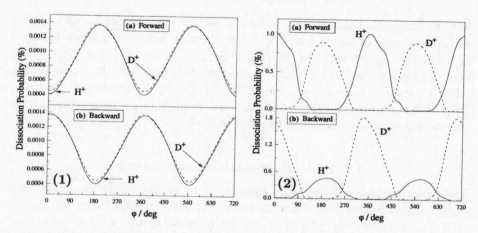

Fig. 4.10. Probabilities of dissociation in the $\{H^+ + D\}$ (solid line) and in the $\{H + D^+\}$ (dashed line) channels in the forward (parts (a)) and backward (parts (b)) directions as a function of the phase shift between the fundamental radiation and its second harmonic. The left-hand part (1) is for the high-frequency domain, with $\lambda_f = 1053$ nm, $\lambda_h = 526.5$ nm and equal intensities $I_f = I_h = 5 \times 10^{12}$ W cm^{-2}; the pulse duration is 150 fs and the initial level $v_0 = 1$. The right-hand part (2) is for the low-frequency domain, with $\lambda_f = 10.6$ μm, $\lambda_h = 5.3$ μm and the intensities $I_f = 5 \times 10^{13}$ W cm^{-2} and $I_h = 2 \times 10^{13}$ W cm^{-2}; the pulse duration is 113 fs and the initial level $v_0 = 0$.

Let us first concentrate on the left-hand part of this figure, in which the following laser parameters have been used: $\lambda_f = 1053$ nm, $\lambda_h = 526.5$ nm with equal intensities $I_f = I_h = 5 \times 10^{12}$ W cm^{-2}. In this 'high-frequency' domain the excitation dynamics is dominated by electronic transitions and consequently almost no distinction between the two hydrogen nuclei H^+ and D^+ is made. The corresponding solid and dashed lines are thus superimposed. However, it is clear in this high-frequency regime that a strong asymmetry in the fragments' angular distributions has been introduced: the phase shift $\varphi = 0$ favours the indirect ejection of the ions in the direction opposite to the detector. The mechanism responsible for this counter-intuitive effect (the maximum of the electric field points towards the detector for $\varphi = 0$, see figure 4.4) will be discussed in the next section. Additionally, the asymmetry in angular distribution is reduced for $\varphi = \pi/2$, in agreement with the electric-field picture of figure 4.4. One can also see that, as explained in section 4.3, adding π to the phase variable φ reverses the angular distributions of ions by sending the ions originally destined for the opposite direction directly to the detector and *vice versa*.

If we now change the wavelengths of the laser pulses to $\lambda_f = 10.6$ μm and $\lambda_h = 5.3$ μm (the 'low-frequency' regime), vibrational transitions play an important role in the molecular excitation. As can be seen on the right-hand side of figure 4.10, a clear distinction between the two isotopes H^+ and D^+ is then introduced. While for $\varphi = 0$, D^+ is still preferentially emitted backwards, H^+ is mainly emitted forwards for the same phase. This separation of isotopes can no longer simply be explained with the help of the classical picture of the electric field since it relies on quantum interferences among different reaction paths.

4.6 Photodissociation experiments

Two-colour phase-control experiments on molecular hydrogen ions have been carried out at two different laser wavelengths and pulse durations. A 50-ps, 1-µm-wavelength field and its second harmonic have been used to dissociate HD^+ and H_2^+ [29], while a shorter-pulse system (85 fs), operating at 750 nm and 375 nm, was used to study photodissociation of H_2^+ [40]. The long-pulse experiments were done at a combined intensity of 2×10^{13} W cm^{-2}, while the short-pulse experiments were done at $\simeq 10^{14}$ W cm^{-2}. In both sets of experiments, the second harmonic was derived from the fundamental by frequency-doubling in a BBO crystal, ensuring phase coherence between the two colours. Their relative phase was varied in the long-pulse experiments by changing the length of dispersive material (glass) traversed by the combined two-colour pulse. In the short-pulse experiments, the two colours were separated in a Michelson interferometer and the relative phase was shifted by varying an arm length, the temporal overlap of the pulses being confirmed by observing the two-colour enhancement of ionisation demonstrated by Shao *et al.* [46].

The two-colour pulse is focused into a chamber containing the sample gas and ion and electron energies are measured using time-of-flight detection. An extraction field directed towards the detector is applied in the interaction region for detection of ions. This permits the distinction between ions with initial velocities directed towards the detector ('forward' ions) and those leaving the interaction region moving away from the detector ('backward' ions). Backward ions are turned around in the field and arrive at the detector with delays determined by their initial energies.

There are several processes occurring in sequence. The neutral molecule is ionised, forming H_2^+ or HD^+. It is subsequently dissociated, forming H^+ + H. During the dissociation, another electron may be removed, forming $H^+ + H^+$. In the HD experiments, the isotopic variants of these products are formed. Previous one-colour studies have shown that the molecular ionisation and the dissociation are sequential and that the ionisation aligns the molecular ion with the linearly polarised laser field [37, 47, 48]. The covariance studies of Frasinski *et al.* [49] and Thompson *et al.* [40]

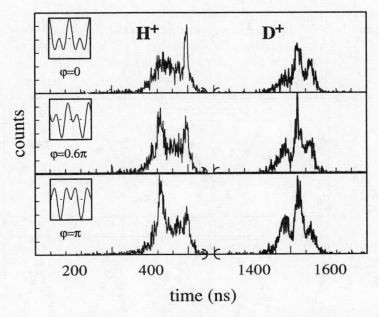

Fig. 4.11. Time-of-flight spectra of H^+ and D^+ ions in the photodissociation of HD^+ by a 1053 nm optical field and its second harmonic. $I_f = I_h = 10^{13}$ W cm^{-2}. The insets show the electric field as a function of time for the specified relative phase φ between the two colours.

demonstrated that the second ionisation stage (forming $H^+ + H^+$) is also sequential. In these studies, it was shown that the formation of $H^+ + H^+$ followed 'Coulomb explosion' dynamics and was responsible for the highest-energy part of the proton spectrum.

Typical time-of-flight spectra for several values of φ, the relative phase between the two colours, are shown in figure 4.11. These spectra were taken in the long-pulse experiment, using HD as the sample gas, at a peak intensity of $I_f = I_h = 10^{13}$ W cm^{-2} and are expanded about the H^+ and D^+ peaks. In each mass peak, ions with zero kinetic energy arrive at the centre of the arrival-time distribution. Forward ions arrive earlier and backward ions arrive later. The D^+ spectrum has an additional central peak due to contaminant H_2 gas (2% of the sample), which is photo-ionised to form H_2^+ with a thermal velocity distribution. The energy at the peak of the distribution is 0.5 eV, which indicates that the predominant mechanism of dissociation is bond-softening [38, 47, 50]. Above-threshold dissociation [38, 47, 50, 51] produces less than 5% of the dissociation yield. There is a clear modulation of the experimental β values, the ratio of forward-ion yield and backward-ion yield, with φ. This is plotted in figure 4.12(a).

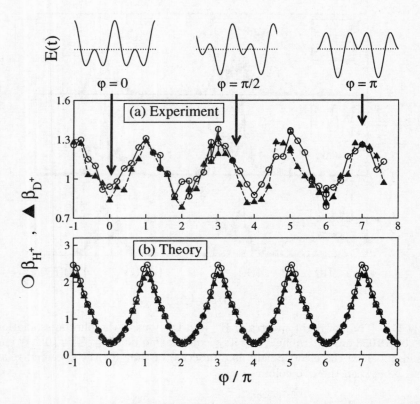

Fig. 4.12. Modulation of the forward/backward-ion-yield ratio β for protons (circles) and deuterons (triangles). (a) Experimental values in the 50-ps and 1053/527-nm photodissociation of HD^+ with $I_f = I_h = 1 \times 10^{13}$ W cm^{-2}, reproduced from [29]. (b) Predicted theoretical values for the same wavelength and intensities with a 0.5-ps pulse, reproduced from [35]. Curves above the graph show the temporal dependence of the electric field at the phases indicated by the arrows.

The periodicity of the forward/backward asymmetry is equal to that of the asymmetry in the two-colour field: the maxima occur at a phase at which the combined electric field is also maximally asymmetric. It is interesting to note, however, that β is maximum when the asymmetric electric-field maximum is pointing *away* from the detector ($\varphi = \pi$ modulo 2π, see figure 4.12), and minimum when the field maximum points *towards* the detector ($\varphi = 0$ modulo 2π). The field is positive when it points towards the detector.

In a naive classical interpretation, one might expect just the opposite: that during the dissociation of the aligned molecular ion, the larger displacement of the electron in the direction opposite to the electric-field

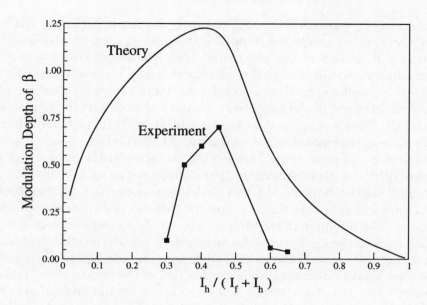

Fig. 4.13. Depth of modulation of the ion ejection asymmetry as a function of the relative intensities in the two colour field. The points are the experimental values, the solid line represents the model of Posthumus *et al.*; reproduced from reference [45].

maximum would tend to leave the nucleus (with a positive charge) on the same side as the field maximum. While this is certainly too simplistic (the dissociation occurs over several optical cycles and, although the electron displacement will be larger opposite to the field maximum, it will also be found on the other side) it seems surprising that the results are precisely the opposite of the prediction.

The direction of emission is also predicted by the theoretical model described in the previous sections. The result of the numerical simulation is plotted in figure 4.12(b). Since the ion-ejection asymmetry arises from the interference between the one- and two-photon couplings between the two electronic states considered, its relationship to the electric-field asymmetry is a function of the frequencies and intensities in the two-colour field. In this 'high-frequency' regime, the theoretical and the experimental results are in good agreement, confirming the counter-intuitive nature of the direction of emission of the ions. It is only under other experimental conditions (the 'low-frequency' regime, see section 4.5) that some ions are ejected in the direction of the field maximum.

In the short-pulse experiments [40], all three dissociation channels, bond-softening, above-threshold dissociation (ATD) and Coulomb explosion, come into play. A phase-dependent forward/backward asymmetry is observed in the H + H$^+$ ATD channel. In this experiment, the ions and electrons were collected simultaneously and again a counter-intuitive direction of ejection of ions was found: both ions and electrons were preferentially ejected opposite to the field maximum. Thompson *et al.* [40, 45] propose a mechanism that is similar to the electron-localisation phenomenon involved in the enhanced ionisation of molecules in intense laser fields [52]. First, the nuclei start to separate, either through bond-softening or by some other mechanism of excitation. At short internuclear distances, the electron can move freely between the nuclei under the influence of the optical field. As the internuclear separation passes a critical value R_c, the potential barrier between the two nuclei rises above the electron-energy level, preventing further transfer, and the electron can be localised on one nucleus. The direction of the electric field as R_c is crossed determines on which nucleus the electron will be localised. While the electric-field maximum at $\varphi = 0$ points in the forward direction, the field is actually pointing in the backward direction for a larger fraction of the optical cycle, making it more likely that the electron will be localised on the forward side and the nucleus moving backwards will exit from the dissociation positively charged.

Using this physical picture, a simplified model has been developed [45]. The electronic wavefunction at an internuclear separation near R_c is approximated as a linear combination of two 1s orbitals, each centred on one of the dissociating nuclei. Using a variational method, the electronic energy in a fixed electric field is minimised and the adiabatic distribution of charge over the two nuclei determined as a function of the instantaneous optical field. The temporal average of this distribution over one cycle of the optical field determines the probabilities of localisation of the electron on each nucleus. This method was used to calculate the intensity dependence of the modulation depth of β (the forward/backward-ion-yield ratio) in the long-pulse experiments on HD$^+$. The results are presented in figure 4.13. The agreement is probably qualitative because the conditions of the model are only partially satisfied in the experiment. The temporal average over an optical cycle in the model requires that the nuclear separation passes through R_c in a time longer than one optical period but, in the experiment, the transit time is closer to half the optical period.

It has been suggested that this model and the theory of the preceding two sections would give different predictions for the ion-ejection asymmetry under some other conditions [45]. However, as yet, no experimental test discriminating between them has been performed.

4.7 Conclusion

This chapter has introduced the basic concepts of the coherent control method and discussed some examples of theoretical as well as experimental implementations in intense laser fields. Clearly, although the mechanisms at the origin of control are much more complex when the interaction between the molecule and the laser field is strong, the high-intensity regime presents a host of new possibilities for controlling the molecular dynamics, with the evident advantage of achieving high yields even with very short pulses.

Beyond the specific scope of the coherent control scheme, a more general approach can also be chosen with the optimal control theory [4]. This approach has proven to be very effective under real experimental conditions [6], using an experimental feedback [5] that automatically guides the modifications of the laser pulses in the quest for an optimal field. This promising approach can undoubtedly also be applied in intense fields, with a high probability of success.

In addition, new methods for the controlled manipulation of individual molecules have been introduced recently [53]. In this scheme, the intensity gradient of a moderately intense laser field has the function of a molecular lens which separates, focuses and transports molecules. These various approaches make it possible to expect a fast development of the control of molecular reactions at the miscroscopic level.

Acknowledgments

E. C. would like to thank Annick Suzor-Weiner and Darian Stibbe for very fruitful comments on the manuscript and acknowledge the CEA for its support from the LRC (grant DSM 97-02). B. S. would like to acknowledge the support of the U.S. Department of Energy at Brookhaven National Laboratory under contract DE-AC02-98CH10886, with the support of the Office of Basic Energy Sciences.

References

[1] I. W. Smith, *Nature* **343** 691 (1990); J. Manz and L. Wöste Eds. *Femtosecond Chemistry* VCH, Weinheim, 1995; A. H. Zewail, *Femtochemistry* World Scientific, Singapore, 1994; M. Chergui Ed. *Femtochemistry* World Scientific, Singapore, 1995.

[2] R. J. Gordon and S. A. Rice, *Annu. Rev. Phys. Chem.* **48** 601 (1997); M. Shapiro and P. Brumer, *J. Chem. Soc. Faraday Trans.* **93** 1263 (1997).

[3] D. J. Tannor and S. A. Rice, *J. Chem. Phys.* **83** 5013 (1985); D. J. Tannor, R. Kosloff and S. A. Rice, *J. Chem. Phys.* **85** 5805 (1986); S. A. Rice, D. J.

Tannor and R. Kosloff, *J. Chem. Soc. Faraday Trans.* **82** 2423 (1986); D. J. Tannor and S. A. Rice, *Adv. Chem. Phys.* **70** 441 (1988).

[4] S. Shi, A. Woody and H. Rabitz, *J. Chem. Phys.* **88** 6870 (1988); A. Peirce, M. Dahleh and H. Rabitz, *Phys. Rev. A* **37** 4950 (1988); R. Kosloff, S. Rice, P. Gaspard, S. Tersigni and D. J. Tannor, *Chem. Phys.* **139** 201 (1989); A. Peirce, M. Dahleh and H. Rabitz, *Phys. Rev. A* **42** 1065 (1990); S. Shi and H. Rabitz, *Comput. Phys. Commun.* **63** 71 (1991); P. Gross, D. Neuhauser and H. Rabitz, *J. Chem. Phys.* **96** 2834 (1992). Y. J. Yan, R. E. Gillilan, R. M. Whitnell, K. R. Wilson and S. Mukamel, *J. Phys. Chem.* **97** 2320 (1993).

[5] R. S. Judson and H. Rabitz, *Phys. Rev. Lett.* **68** 1500 (1992); B. Amstrup, G. J. Toth, G. Szabo, H. Rabitz and A. Lörincz, *J. Phys. Chem.* **99** 5206 (1995).

[6] C. J. Bardeen, V. V. Yakovlev, K. R. Wilson, S. D. Carpenter, P. M. Weber and W. S. Warren, *Chem. Phys. Lett.* **280** 151 (1997); T. Baumert, T. Brixner, V. Seyfried, M. Strehle and G. Gerber, *Appl. Phys. B* **65** 779 (1997); A. Assion, T. Baumert, M. Bergt, T. Brixner, B. Kiefer, V. Seyfried, M. Strehle and G. Gerber, *Science* **282** 919 (1998).

[7] T. Baumert and G. Gerber, *Isr. J. Chem.* **34** 103 (1994).

[8] V. Blanchet, M. A. Bouchene, O. Cabrol and B. Girard, *Chem. Phys. Lett.* **233** 491 (1995); V. Blanchet, C. Nicole, M. A. Bouchene and B. Girard, *Phys. Rev. Lett.* **78** 2716 (1997); M. A. Bouchene, V. Blanchet, C. Nicole, N. Melikechi, B. Girard, H. Ruppe, S. Rutz, E. Schreiber and L. Woeste, *Eur. Phys. J. D* **2** 131 (1998).

[9] B. Kohler, V. V. Yakovlev, J. Che, J. L. Krause, M. Messina, K. R. Wilson, N. Schwentner, R. M. Whitnell and Y. Yan, *Phys. Rev. Lett.* **74** 3360 (1995); A. Assion, T. Baumert, J. Helbing, V. Seyfried and G. Gerber, *Chem. Phys. Lett.* **259** 488 (1996); J. J. Baumberg, A. P. Heberle, K. Kohler and K. Ploog, *J. Opt. Soc. Am. B* **13** 1246 (1996); N. H. Bonadeo, J. Erland, D. Gammon, D. Park, D. S. Katzer and D. G. Steel, *Science* **282** 1473 (1998); D. Meshulach, D. Yelin and Y. Silberberg, *J. Opt. Soc. Am. B* **15** 1615 (1998); V. V. Lozovoy, S. A. Antipin, F. E. Gostev, A. A. Titov, D. G. Tovbin, O. M. Sarkisov, A. S. Vetchinkin and S. Y. Umanskii, *Chem. Phys. Lett.* **284** 221 (1998); T. C. Weinacht and P. H. Bucksbaum, *Nature* **397** 233 (1999).

[10] M. Machholm and A. Suzor-Weiner, *J. Chem. Phys.* **105** 971 (1996).

[11] P. Brumer and M. Shapiro, *Chem. Phys.* **139** 221 (1989).

[12] Z. Chen, M. Shapiro and P. Brumer, *Chem. Phys. Lett.* **228** 289 (1994); Z. Chen, M. Shapiro and P. Brumer, *J. Chem. Phys.* **102** 5683 (1995); Z. Chen, M. Shapiro and P. Brumer, *Phys. Rev. A* **52** 2225 (1995).

[13] A. Schnitman, I. Sofer, I. Golub, A. Yogev, M. Shapiro, Z. Chen and P. Brumer, *Phys. Rev. Lett.* **76** 2886 (1996).

[14] P. Brumer and M. Shapiro, *Faraday Discuss. Chem. Soc.* **82** 177 (1986); P. Brumer and M. Shapiro, *Chem. Phys. Lett.* **126** 541 (1986); M. Shapiro and P. Brumer, *J. Chem. Phys.* **84** 4103 (1986); M. Shapiro, J. Hepburn and P. Brumer, *Chem. Phys. Lett.* **149** 451 (1988); P. Brumer and M. Shapiro, *Acc. Chem. Res.* **22** 407 (1989); P. Brumer and M. Shapiro, *Annu. Rev. Phys. Chem.* **43** 257 (1992).

[15] P. Brumer and M. Shapiro, in *Molecules in Laser Fields* A. D. Bandrauk Ed, New York, Dekker, 1994.

[16] N. F. Scherer, R. J. Carlson, A. Matro, M. Du, A. J. Ruggiero, V. Romero-Rochin, J. A. Cina, G. R. Fleming and S. A. Rice, *J. Chem. Phys.* **95** 1487 (1991).

[17] C. Chen, Y. Y. Yin and D. S. Elliot, *Phys. Rev. Lett.* **64** 507 (1990); C. Chen and D. S. Elliot, *Phys. Rev. Lett.* **65** 1737 (1990); N. E. Karapanagioti, D. Xenakis, D. Charalambidis and C. Fotakis, *J. Phys. B* **29** 3599 (1996); S. Cavalieri, R. Eramo and L. Fini, *Phys. Rev. A* **55** 2941 (1997).

[18] D. W. Schumacher, F. Weihe, H. G. Muller and P. H. Bucksbaum, *Phys. Rev. Lett.* **73** 1344 (1994).

[19] S. M. Park, S. P. Lu and R. J. Gordon, *J. Chem. Phys.* **94** 8622 (1991); S. P. Lu, S. M. Park, Y. Xie and R. J. Gordon, *J. Chem. Phys.* **96** 6613 (1992).

[20] V. Kleiman, L. Zhu, X. Li and R. Gordon, *J. Chem. Phys.* **102** 5863 (1995); W. Xuebin, R. Bersohn, K. Takahashi, M. Kawasaki and K. L. Hong, *J. Chem. Phys.* **105** 2992 (1996).

[21] G. Xing, X. Wang, X. Huang, R. Bersohn and B. Katz, *J. Chem. Phys.* **104** 826 (1996).

[22] V. D. Kleiman, L. Zhu, J. Allen and R. J. Gordon, *J. Chem. Phys.* **103** 10 800 (1995).

[23] L. Zhu, V. D. Kleiman, X. Li, S. P. Lu, K. Trentelman and R. J. Gordon, *Science* **270** 77 (1995); L. Zhu, K. Suto, J. A. Fiss, R. Wada, T. Seideman and R. J. Gordon, *Phys. Rev. Lett.* **79** 4108 (1997); J. A. Fiss, L. Zhu, K. Suto, G. He and R. J. Gordon, *Chem. Phys.* **233** 335 (1998); J. A. Fiss, L. Zhu, R. J. Gordon and T. Seideman, *Phys. Rev. Lett.* **82** 65 (1999).

[24] G. Kurizki, M. Shapiro and P. Brumer, *Phys. Rev. B* **39** 3435 (1989).

[25] N. B. Baranova, A. N. Chudinov and B. Y. Zel'dovich, *Opt. Commun.* **79** 116 (1990); E. Dupont, P. B. Corkum, H. C. Liu, H. C. Liu, M. Buchanan and Z. R. Wasilewski, *Phys. Rev. Lett.* **74** 3596 (1995); R. Atanasov, A. Hache, J. L. P. Hughes, H. M. van-Driel and J. E. Sipe, *Phys. Rev. Lett.* **76** 1703 (1996).

[26] H. G. Muller, P. H. Bucksbaum, D. W. Schumacher and A. Zavriyev, *J. Phys. B* **23** 2761 (1990); A. Szöke, K. C. Kulander, and J. N. Bardsley, *J. Phys. B* **24** 3165 (1991); K. J. Schafer and K. C. Kulander, *Phys. Rev. A* **45** 8026 (1992); E. E. Aubanel and A. D. Bandrauk, *Chem. Phys. Lett.*

229 169 (1994); A. D. Bandrauk and E. E. Aubanel, *Chem. Phys.* **198** 159 (1995).

[27] R. M. Potvliege and P. H. G. Smith, *J. Phys. B* **24** L641 (1991); R. M. Potvliege and P. H. G. Smith, *J. Phys. B* **25** 2501 (1992); R. M. Potvliege and P. H. G. Smith, *Phys. Rev. A* **49** 3110 (1994).

[28] Y. Y. Yin, C. Chen and D. S. Elliot, *Phys. Rev. Lett.* **69** 2353 (1992); K. L. Hong and R. Bersohn, *J. Chem. Phys.* **107** 4546 (1997).

[29] B. Sheehy, B. Walker and L. F. DiMauro, *Phys. Rev. Lett.* **74** 4799 (1995).

[30] M. J. J. Vrakking and S. Stolte, *Chem. Phys. Lett.* **271** 209 (1997).

[31] T. Nakajima and P. Lambropoulos, *Phys. Rev. A* **50** 595 (1994).

[32] M. Protopapas, A. Sanpera, P. L. Knight and K. Burnett, *Phys. Rev. A* **52** R2527 (1995); A. D. Bandrauk, S. Chelkowski, H. Yu and E. Constant, *Phys. Rev. A* **56** R2537 (1997); I. C. Shih and M. T. Xiao, *Int. J. Quant. Chem.* **69** 293 (1998).

[33] M. Protopapas, P. L. Knight and K. Burnett, *Phys. Rev. A* **49** 1945 (1994).

[34] R. A. Blank and M. Shapiro, *Phys. Rev. A* **52** 4278 (1995); R. Taieb, V. Veniard and A. Maquet, *J. Opt. Soc. Am. B* **13** 363 (1996); T. Zuo and A. D. Bandrauk, *Phys. Rev. A* **54** 3254 (1996).

[35] E. Charron, A. Giusti-Suzor and F. H. Mies, *J. Chem. Phys.* **103** 7359 (1995).

[36] E. Charron, A. Giusti-Suzor and F. H. Mies, *Phys. Rev. Lett.* **71** 692 (1993); E. Charron, A. Giusti-Suzor and F. H. Mies, *Phys. Rev. Lett.* **75** 2815 (1995).

[37] E. Charron, A. Giusti-Suzor and F. H. Mies, *Phys. Rev. A* **49** R641 (1994).

[38] A. Giusti-Suzor, F. H. Mies, L. F. DiMauro, E. Charron, and B. Yang, (Topical review) *J. Phys. B* **28** 309 (1995).

[39] Eric Charron, *Thèse de Doctorat* Université Paris XI, 1994. Available on http://www.ppm.u-psud.fr

[40] M. R. Thompson, M. K. Thomas, P. F. Taday, J. H. Posthumus, A. J. Langley, L. J. Frasinski and K. Codling, *J. Phys. B* **30** 5755 (1997).

[41] G. A. Edgar, *Measure Topology and Fractal Geometry* Springer-Verlag, 1990; L. S. Liebovitch and T. Toth, *Phys. Lett. A* **141** 386 (1989).

[42] C. Leforestier, H. Bisseling, C. Cerjan, M. D. Feit, R. Friesner, A. Guldberg, A. Hammerich, G. Jolicard, W. Karrlein, H. D. Meyer, N. Lipkin, O. Roncero and R. Kosloff, *J. Comput. Phys.* **94** 59 (1991).

[43] M. J. Feit, J. A. Fleck and A. Steiger, *J. Comput. Phys.* **47** 412 (1982).

[44] *Molecules in Laser Fields* A. D. Bandrauk Ed., New York, Dekker, 1994.

[45] J. H. Posthumus, M. R. Thompson, A. J. Giles and K. Codling, *Phys. Rev. A* **54** 955 (1996).

[46] Y. L. Shao, D. J. , M. H. R. Hutchinson, J. Larsson, J. P. Marangos and J. W. G. Tisch, *J. Phys. B* **29** 5421 (1996).

[47] P. H. Bucksbaum, A. Zavriyev, H. G. Muller and D. W. Schumacher, *Phys. Rev. Lett.* **64** 1883 (1990); A. Zavriyev, P. H. Bucksbaum, H. G. Muller and D. W. Schumacher, *Phys. Rev. A* **42** 5500 (1990); B. Yang, M. Saeed, L. F. DiMauro, A. Zavriyev and P. H. Bucksbaum, *Phys. Rev. A* **44** R1458 (1991).

[48] D. Normand, L. A. Lompre and C. Cornaggia, *J. Phys. B* **25** L497 (1992); D. T. Strickland, Y. Beaudouin, P. Dietrich and P. B. Corkum, *Phys. Rev. Lett.* **68** 2755 (1992).

[49] L. J. Frasinski, M. Stankiewicz, P. A. Hatherly, G. M. Cross, K. Codling, A. J. Langley and W. Shaikh, *Phys. Rev. A* **46** R6789 (1992).

[50] K. Codling and L. J. Frasinski, *J. Phys. B* **26** 783 (1993).

[51] A. Giusti-Suzor, X. He, O. Atabek and F. H. Mies, *Phys. Rev. Lett.* **64** 515 (1990).

[52] M. Schmidt, D. Normand and C. Cornaggia, *Phys. Rev. A* **50** 5037 (1994); T. Seideman, M. Ivanov and P. B. Corkum, *Phys. Rev. Lett.* **75** 2819 (1995); J. H. Posthumus, L. J. Frasinski, A. J. Giles and K. Codling, *J. Phys. B* **28** L349 (1995).

[53] T. Seideman, *J. Chem. Phys.* **106** 2881 (1997); T. Seideman, *Phys. Rev. A* **56** R17 (1997); T. Seideman, *J. Chem. Phys.* **107** 10 420 (1997); T. Seideman and V. Kharchenko, *J. Chem. Phys.* **108** 6272 (1998).

5

Experimental studies of laser-heated rare-gas clusters

Matthias Lezius[†] and Martin Schmidt

*CEA-DSM/DRECAM/SPAM, C. E. de Saclay, Bâtiment 524,
91191 Gif-sur-Yvette Cedex, France*

[†] *Present address: Universität Innsbruck,
Institut für Ionenphysik, Technikerstraße 25, 6020 Innsbruck, Austria*

5.1 Introduction

The advent of compact, high-power lasers in the last two decades has opened a rich and challenging research field, aimed at understanding the behaviour of matter under the influence of ultra-intense and extremely short pulses of light. Typical pulse durations of less than 10^{-13} s are now available with commercial laser systems that are designed to deliver peak powers beyond 10^{12} W. The focusing of such laser pulses leads to extremely high energy intensities that may exceed 10^{18} W cm^{-2}. The corresponding electric-field strengths are of the order of several GV cm^{-1} and megagauss magnetic fields. In terms of photon numbers, at 10^{17} W cm^{-2} the irradiation is already so intense that approximately 10 000 photons will pass through the volume of one atom during each optical cycle (3 fs).

In the last few decades, extensive strong-laser-field studies have been performed on atomic targets in the gas phase [1] as well as on bulk solid-state targets [2]. Several pertinent phenomena in strong fields were discovered as a result of the non-linear response of matter to intense laser irradiation. For the particular case of atoms in the gas phase, multi-photon ionisation (MPI) and optical field ionisation (OFI) have been studied in great detail both experimentally [3–6] and theoretically [7–11]. Moreover, the observation of highly non-linear phenomena such as above-threshold ionisation (ATI) [12, 13] and high-order-harmonic generation (HHG) [14–18] and their detailed theoretical description [19] have lead to a comprehensive understanding of the principal physical processes that govern the atomic response to intense laser fields.

The studies have been extended to diatomic molecules [20–22], see chapter 2, as well as polyatomic [23] molecules, see chapter 3, in order to investigate the influence of chemical bonds that are induced by exchange

142

forces of the valence electrons. For example, it was observed that interaction with a strong field provokes an alignment of the molecule along
the electric-field vector of the laser beam, then multiple ionisation of
several valence electrons and finally the rupture of the chemical system
by Coulomb repulsion of the nuclei. This phenomenon, usually denoted
multi-electron dissociative ionisation (MEDI), leads to the confined emission of multiply charged atomic fragments. Usually, quite modest kinetic
energies of the order of a few tens of electron-volts are observed.

Many concepts that were developed for atoms still apply to molecules
[24]. However, a typical result, which arises specifically from the molecular
structure, was the discovery of the so-called critical distance at which multiple ionisation preferentially occurs [25, 26]. In the simplest, diatomic,
case this new property of molecules is related to OFI of electrons from
the double potential well. Assuming that a simple point-charge model
applies, the critical distance can explain the observed releases of kinetic
energy. An enhancement of ionisation at a critical separation of the nuclei is the direct result of the influence of the adjacent ions. The situation
leads to a significant lowering of the (internal) tunnel-ionisation barrier
[26, 27].

Since the beginning of the development of high-performance laser systems also bulk targets have been subjected to ever more intense and
shorter pulses of light. This type of interaction leads to the production of
very hot plasma as a result of phenomena with a distinctly collective character, e.g. electron–ion collisional heating and hydrodynamic expansion
[28]. Generally, such laser-produced plasmas generate copious amounts of
X-rays [29] and are a source of hot electrons and energetic, highly charged
ions [30]. These features are in clear contrast to the strong-field behaviour
of low-density atomic gas-phase targets. A rich literature on the complex
physics of laser-produced plasma is now available and significant progress
towards the understanding of the underlying collective processes has been
made.

In many areas of research *clusters* attract the attention of chemists and
physicists, since they represent an intermediate state of matter, bridg-
ing the gap between atomic and solid-state targets [31]. Most of these
studies concerned the evolution of the physical and chemical properties
of materials as they change progressively from the monatomic to the bulk
structure. Surprisingly, despite such a unique role of gas-phase cluster targets having been appreciated, strong-field studies started only a few years
ago. Prominent examples are the experimental studies on generation of
X-rays from rare-gas clusters by Rhodes *et al.* [32] and from fullerenes by
Chichkov and co-workers [33] and of Falcone and co-workers [34], who used
gold clusters. The multiple ionisation and dissociation of mixed molecular
clusters was first studied by Castleman *et al.* [35]. Later, in 1996, Snyder

et al. [36] reported the production of 20-times-charged atomic ions after irradiation of neat xenon clusters with moderately intense laser fields of 10^{15} W cm^{-2}. As a last example, Ditmire *et al.* [37] produced ions up to Xe^{30+} with very high kinetic energies, approaching 1 MeV, from laser-heated xenon clusters by applying even more intense and shorter laser pulses.

These experimental results strongly suggest that small-scale collective effects are involved. Very clearly, clusters exhibit features similar to those normally observed for solid targets, i.e. intense emission of X-rays and the production of highly charged ions. Several models have been proposed thus far in an attempt to explain the experimental observations. For example, Rhodes and co-workers [32] proposed the *coherent-electron-motion* model, in which several electrons collide coherently with the ion cores within a cluster, thereby leading to efficient ionisation of inner-shell electrons at very high laser intensities. Wulker *et al.* [33] were the first to propose a *collisional-absorption* mechanism for the laser-induced heating of clusters. Perry and co-workers [38] presented sophisticated numerical simulations based on the same mechanism for argon clusters, together with experimental data. Finally, Rose-Petruck *et al.* [39] performed Monte Carlo simulations that demonstrated the possibility of *ignition-like enhancement of ionisation* in complex, finite systems, due to a lowering of the OFI barrier by adjacent, highly charged ions, i.e. by a process analogous to that governing multiple ionisation of molecules [26, 27]. In fact, all these models include rescattering of the liberated electrons within the cluster, since OFI alone cannot explain the intense response to relatively modest laser fields. Even after accounting for the enhancement of the electric field due to the different refractive index of the ionised cluster medium, without rescattering the laser intensity would have to be orders of magnitude higher.

A characteristic feature of clusters is their size, which is generally smaller than the wavelength of the incident laser beam and the typical skin depth of solid targets. This is sometimes believed to lead to a very efficient coupling between light and matter. In such a case clusters would be a potential source for the production of very intense and ultra-short bursts of X-rays. In this sense it should be noted that clusters of noble gas atoms are sustainable and debris-free targets. Ultra-short pulsed X-ray sources could be of considerable interest for various applications in material processing, medicine and biology. In this context, X-ray sources driven by so-called table-top terawatt lasers may become competitive against conventional sources such as synchrotrons, particularly in terms of cost and user-friendliness.

Since the beginning of 1996 the Saclay group has joined the international efforts investigating the fascinating domain of laser–cluster interactions. Our principal goal was to understand the explosion dynamics of neat rare-gas clusters and to measure the kinetic energies of the fragment ions as a function of the size and constituents of the cluster. We have also performed systematic experiments in order to provide the first absolute measurements of efficiency of production of X-rays with cluster-jet targets.

In section 5.2 we will present a brief introduction of cluster-production techniques with emphasis on the synthesis of rather large aggregates of rare-gas atoms. We will discuss the standard techniques for the estimation of mean cluster sizes which make use of several well-established scaling laws. In section 5.3 we will summarise some of the fundamental physical laws that are involved in the description of the laser–cluster interaction. To this end we use a collisional absorption model, which is in line with the one first developed by Perry and co-workers [38]. There, we will also briefly discuss our current interpretation of the kinetic energies of ions. In section 5.4 we present the interesting case of large noble-gas clusters that are irradiated by comparatively weak (10^{14} W cm^{-2}) laser pulses of duration 30 ps. This situation leads to a typical plasma-like expansion following more or less conventional hydrodynamic laws. In section 5.5 this can be compared with the experimental results obtained after irradiation with 130-fs pulses at intensities up to 5×10^{17} W cm^{-2}. Such an appreciably higher intensity of irradiation leads to a number of non-linear effects and to a very distinct outcome of the laser–matter interaction. Finally, in section 5.6 we present X-ray spectra and quantitative measurements of the emission patterns and photon-conversion efficiencies for rare-gas clusters. These X-ray data are particularly useful for obtaining information on the ionisation states reached during the short-pulse irradiation. Moreover, this work contributes to the international efforts at developing and characterising short-pulsed X-ray sources.

5.2 Cluster sources

Since the first observation of clusters [40], the research related to these nanometre-sized aggregates has greatly stimulated the scientific community and produced important results such as the discovery of carbon clusters [41]. It is beyond the scope of this article to review the various techniques that have been employed so far for the production of atomic, molecular, metal and salt clusters. A recent survey on cluster sources can be found in the centennial issue of the *Journal of Physical Chemistry* [42]; earlier reviews were given by Märk and Castleman [43].

For the special case of investigating nanoplasma behaviour it is sufficient to produce rather large neutral clusters in sufficient quantities. Generally, these clusters are easily obtained by expanding rare gases or molecular gases through a nozzle into a vacuum. It is worth mentioning that, from our current understanding, large metal clusters should also become an important target for intense laser irradiation. Metal clusters can in principle be produced by a laser-evaporation source inside an adiabatically expanding, inert carrier gas, e.g. helium. Since these sources behave similarly to a pure gas expansion, the efficiency of production of metal clusters follows similar rules [44]. Therefore, despite the fact that there are numerous methods of production, we restrict our discussion to the formation of neutral, medium-to-large-sized atomic clusters during adiabatic expansion.

A particular difficulty of cluster-jet studies resides in the fact that the target has a rather inhomogeneous character, i.e. the production of clusters in an adiabatic expansion always results in a broad distribution of cluster sizes and, in addition, with pulsed jets the concentration of clusters varies both in time and in space. Unlike properties of low-density atomic and solid targets, the properties of cluster targets are not easily characterised, which makes it very difficult to obtain reproducible results. We have based our work on the investigations by Hagena [45] on corresponding nozzle flows. These papers provide a collection of semi-empirical scaling laws that we briefly summarise here.

The maximum flow velocity for an adiabatic expansion of a stagnant gas with temperature T_0 is given by

$$v_\infty = 1.581 \sqrt{\frac{2k_B T_0}{m}}, \tag{5.1}$$

with m being the molecular mass in atomic units. At a sufficiently large distance, x, downstream of the nozzle, the particle density n and the temperature T in the jet are given by

$$n(\delta)/n_0 = 0.1\delta^{-1}; \ T(\delta)/T_0 = 0.282\delta^{-\frac{4}{3}}, \tag{5.2}$$

where $\delta = x/d$, the ratio of x and the nozzle diameter. The input parameters of the backing gas are its density, n_0, and its temperature, T_0. At a distance x_∞, the expansion passes into the collisionless-flow regime and the terminal temperature T_∞ is reached. T_∞ can be estimated using the hard-sphere approximation:

$$T_\infty \approx K(n_0 d T_0^{-1.25})^{-0.8}, \tag{5.3}$$

where K is a gas-specific, empirical constant. In physical terms, clustering occurs when the expanding and cooling gas becomes super-saturated, i.e. when the line of adiabatic expansion crosses the vapour-pressure curve. Unfortunately, there is still no complete theory that can describe the onset and growth of clustering. Nevertheless, the final degree of clustering can be predicted with the empirical scaling parameter

$$\Gamma = n_0 d^q T_0^{0.25q-1.5} \; (q \approx 0.85). \tag{5.4}$$

For rare gases, this concept, which is based on corresponding nozzle flows, is in excellent agreement with experiment. It can be extended to metal clusters by using the following characteristic values for length and energy:

$$r_{ch} = (m/\rho)^{\frac{1}{3}}; \; T_{ch} = \Delta h_0^\circ / k_B, \tag{5.5}$$

where ρ is the density of the solid and Δh_0° is the enthalpy of sublimation [44]. With the help of these characteristic values Γ can be transformed into the reduced scaling parameter

$$\Gamma^* = \frac{\Gamma}{r_{ch}^{q-3} T_{ch}^{0.25q-1.5}} = K_\Gamma P_0 d^{0.85} T_0^{-2.29}. \tag{5.6}$$

Note that Γ^* depends only linearly on the stagnant gas pressure P_0, while the stagnant temperature T_0 has a much stronger effect. The empirical values for K_Γ, for a variety of gases and metals, are listed in [46]. Flows created with the same Γ^* will be similar with respect to cluster formation. The average cluster size $\bar{N}$ can be estimated from

$$\bar{N} = 33(\Gamma^*/1000)^{1.95}. \tag{5.7}$$

The efficiency of cluster production is significantly enhanced by the use of a conical nozzle, which is characterised by the cone angle 2α. The flow through a small-angle, conical nozzle is much smaller than that through a sonic nozzle with the same efficiency of cluster production. The equivalent size for a sonic nozzle is given by

$$d_{eq}/d = 0.736/\tan \alpha. \tag{5.8}$$

For example, by using a cone angle of $10°$, it is possible to enhance the efficiency of cluster production by a factor of 8.4, whilst reducing the gas flow by a factor of 70.

Well-collimated cluster beams may have high particle densities of between 10^{14} and 10^{24} molecules per m^3 [47]. Thus the large gas flow, which is needed for obtaining the necessary high cooling rates during the adiabatic expansion, leads to a poor vacuum near the expansion zone. Keeping in mind that the intense laser pulse will ionise with very high efficiency at the focus, it is advantageous to reduce the amount of background gas by using one or more stages of skimmed, differential pumping. Otherwise, space-charge and micro-plasma effects can easily contaminate the experimental results. In addition, the distance between the first molecular-beam skimmer and the nozzle has to be chosen carefully in order to minimise interference between the Mach disc and the orifice of the skimmer. Such an interference can significantly disturb the quality and stability of the beam of clusters. As a rule of thumb, for an initial gas pressure P_0 and a background pressure P_1, the axial position x_M of the Mach disc is given by

$$x_M/d = 0.65\sqrt{P_0/P_1}. \tag{5.9}$$

With $P_0 \approx 30$ bar and a conical nozzle of dimensions 300 μm and 10°, we observed the highest cluster intensities for nozzle–skimmer distances exceeding 70 mm.

Finally, since ultra-intense lasers are generally pulsed, with repetition rates between 0.01 Hz and 1 kHz, it is beneficial to reduce the overall gas flow of the supersonic expansion by using pulsed nozzle valves. This reduces the cost of the vacuum system significantly. Pulsed nozzle sources with high repetition rates and well-defined opening functions are commercially available or can be self-assembled by following descriptions like the one in [48].

5.3 The interaction of intense lasers with clusters

The interaction of intense lasers with clusters is very different from that with small molecules. There are two main reasons for this. Firstly, clusters absorb energy from the electro-magnetic field by a different mechanism from that for small molecules. Secondly, depending on the size of cluster, the decay of the system will be dominated either by the electron temperature or by the local space charge.

We will briefly describe the absorption of energy by clusters in connection with the mechanisms for atoms and small molecules. From numerous studies it is known that the atomic and molecular responses are entirely determined by multiple-ionisation processes based either on OFI or on MPI (cf. section 5.1). For molecules, the removal of several valence electrons leads to the rupture of the chemical bond and the multiply charged fragments experience Coulomb repulsion. In the case of clusters, however, the constituents are much more efficiently ionised due to collective intra-cluster rescattering of the electrons that were initially produced by OFI. In the locally dense cluster medium the mean free path of the electrons is estimated from the standard Spitzer formula:

$$\lambda_e = \frac{(k_B T_e)^2}{4\pi n_e (Z+1)e^4 \ln \Lambda}, \tag{5.10}$$

with

$$\ln \Lambda = \min\left\{\ln\left(\frac{2.58 \times 10^{12} T_{\mathrm{e}}^{1.5}}{Z\sqrt{n_{\mathrm{e}}}}\right); \ln\left(\frac{2.83 \times 10^{13} T_{\mathrm{e}}^{1.5}}{\sqrt{n_{\mathrm{e}}}}\right)\right\}. \qquad (5.11)$$

Here $\ln \Lambda$ denotes the Coulomb logarithm, T_{e} the electron temperature, n_{e} the electron density and Z the average charge state of the ions. However, this is only a first-order approximation since the strong electro-magnetic field will additionally accelerate the electrons between collisions. A more careful approach was proposed by Ditmire *et al.* [38]. They used numerical integration of the formulas of Silin [49], which, in addition, must be limited by a suitably chosen maximum for the collision frequency ν_{max} at the electric field E:

$$\nu_{\mathrm{max}} \approx \frac{2eEn_{\mathrm{i}}^{\frac{1}{3}}}{\pi m_{\mathrm{e}}\omega_0}. \qquad (5.12)$$

Typical maximum collision frequencies for ion densities n_{i} reaching solid densities can be of the order of the laser frequency ω_0. This implies that collisional heating is the dominating process as long as the clusters are large enough to make rescattering probable.

For electron–ion collisional absorption, the inverse of the absorption length K can be obtained from the theory for inverse *Bremsstrahlung* published by Dawson and Johnston [50]:

$$K = \frac{Z^2 n_{\mathrm{e}} n_{\mathrm{i}}}{3\epsilon_0 c \omega_0} \frac{\ln \Lambda}{(2\pi m_{\mathrm{e}} k_{\mathrm{B}} T_{\mathrm{e}})^{\frac{3}{2}}} \frac{1}{\sqrt{1 - \omega_{\mathrm{p}}^2/\omega_0^2}}, \qquad (5.13)$$

where $\omega_{\mathrm{p}} = \sqrt{4\pi e^2 n_{\mathrm{e}}/m_{\mathrm{e}}}$ is the plasma frequency. Additionally, the heating rate for collisional absorption into clusters can be described by considering the rate of deposition of laser energy into a dielectric sphere [38]:

$$\frac{\partial U}{\partial t} = \frac{9\omega \, \Im(\epsilon)}{|\epsilon + 2|^2} E_0^2. \qquad (5.14)$$

The Drude model can be used to calculate the dielectric constant of the plasma (see e.g. [51]):

$$\epsilon = 1 - \frac{\omega_{\mathrm{p}}^2}{\omega(\omega + i\nu)}. \qquad (5.15)$$

It is important to note that this approach represents quite a severe approximation, since the laser-heated cluster very quickly becomes a conducting sphere, due to the high degree of ionisation of the constituents. However, the great difficulties in obtaining a more correct description of the process of absorption make the above approximation defensible.

As an important result of the model, the efficiency of heating appears to become resonantly enhanced when $n_e = 3m_e\omega^2/(4\pi e^2)$. For example, for argon clusters with a van der Waals equilibrium distance, resonant heating should start to occur for an average degree of ionisation of less than 2. From the more general equation (5.13), it is already clear that the absorption will tune into resonance when $\omega_p = \omega_0$, a situation in which the plasma reaches the so-called critical density n_c. Interestingly, the free-electron density of an unexpanded cluster with $Z = 1$ exceeds n_c. Therefore, all expanding cluster systems should be driven through resonant absorption somewhere between n_c and $3n_c$.

An important difficulty in describing decaying irradiated systems is that most of the parameters governing collisional absorption (equation (5.13)) are time-dependent, namely $Z(t)$, $n_e(t)$, $n_i(t)$, $T_e(t)$, $\omega_p(t)$ and $\ln \Lambda(t)$. This makes simulations extremely difficult and sensitive to the boundary conditions [52]. Thus far the only self-consistent simulation has been that performed by Ditmire et al. [38]. From their calculations the authors find a short period (approximately 20 fs) of resonantly enhanced electron heating, approximately 100 fs after the beginning of a 130-fs laser pulse of 2×10^{16} W cm^{-2} peak intensity. For lower intensities the resonance shifts to longer time delays and becomes significantly broadened. The simulation demonstrates that resonant absorption of light begins as soon as the cluster's radius has increased by a factor of three. The subsequent, rapid increase of the energy density leads to an explosion-like expansion of the aggregate.

An intensely laser-heated cluster behaves like a very small, spherical plasma; we will refer to such a physical entity as a nanoplasma in the following. A plasma is generally considered globally neutral; this applies for example to a macroscopic plasma obtained from solid targets. There are several sophisticated theories of the expansion dynamics of hot, neutral plasmas of spherical symmetry. For example, in the model proposed by Haught and Polk [53], the plasma is considered as a fluid. The dynamics of the system is solely determined by the pressure of the hot electrons, leading to ambipolar expansion of the plasma as a whole. This so-called hydrodynamic pressure is given by the expression

$$P_{\mathrm{H}} = n_e k_{\mathrm{B}} T_e. \qquad (5.16)$$

During expansion, the density obviously drops proportionally to $1/r^3$, with r the radius of the plasma ball. The hydrodynamic pressure P_{H} will therefore also decrease as $1/r^3$, since, according to the self-similar model of [53], the temperature is only a function of time. The evolution of the diameter of a plasma consisting of ions with mass m_i and density n_i is described by the differential equation [53]

$$\frac{\partial^2 r}{\partial t^2} = \frac{3P_H}{n_i m_i r}.$$ (5.17)

However, when one is considering a *nanoplasma*, it is necessary to take the loss of numerous electrons, as a result of the laser-induced heating, into account.

These so-called free-streaming electrons will lead to a positive excess charge Q of the nanoplasma, thereby producing an additional pressure, denoted the Coulomb pressure:

$$P_{Cb} = \frac{3Q^2 e^2}{2(4\pi)^2 \epsilon_0 r^4}.$$ (5.18)

The cluster-explosion dynamics will therefore be characterised by the relative contributions of P_H and P_{Cb}. From the above considerations one would expect that the Coulomb pressure prevails for smaller clusters, whilst the hydrodynamic pressure mainly drives the expansion of a larger system. We have studied such cluster-size effects in detail and present the results of our studies in section 5.5.

Another important parameter that strongly influences the cluster-explosion dynamics is the duration of the laser pulse. When one is using ultra-short laser pulses, the collisional heating stops before the cluster has blown up to a significant size. Such a situation of inertial confinement arises particularly for clusters consisting of high-Z constituents. Important works regarding effects of the pulse duration have been published recently by Ditmire and coworkers [54]. It should be noted that the description of the heating process becomes increasingly complex when the laser pulse lasts much longer than the typical time taken for expansion of a cluster to the critical density n_c. For example, during irradiation with a laser pulse of duration a few picoseconds, the cluster will expand to several times its original diameter. However, if the laser intensity is chosen low enough, such that only a small number of electrons can leave the cluster, the Coulomb pressure should play a negligible role. We will see in section 5.4 that this is indeed the case and that the explosion of such a system is mainly governed by hydrodynamic forces.

Experimental details

Most of our experiments were carried out at the Ti:sapphire femtosecond laser facility of the DRECAM at the CEA Saclay. This system can deliver 130-fs pulses with a maximum energy of 80 mJ, at a wavelength of 790 nm and at a repetition rate of 20 Hz. Focusing of the pulses by means of an off-axis parabolic mirror yielded peak intensities up to 5×10^{17} W cm^{-2}.

Occasionally, we also used a mode-locked Nd: YAG laser, delivering 30-ps pulses of 50 mJ at 1064 nm. Focusing with the same mirror allowed us to produce maximum intensities of 2×10^{14} W cm^{-2}.

For the analysis of the fragment ions and electrons, we used a time of flight (TOF) mass spectrometer that had previously been employed in studies of molecules in strong fields [25]. The simultaneous detection of ions, electrons and photons was realised with a particle detector positioned approximately 30 cm from the laser focus. The various particles were detected at different delay times with respect to the incoming laser pulse. In experiments in which electrons, ions and photons were analysed simultaneously, an electron multiplier with an earthed entrance dynode was used for the detection of particles. For the analysis of the fast, multiply charged fragment ions, we used a time-of-flight spectrometer with magnetic deflection (MD-TOF) [55]. This is similar to a single-shot device recently developed by Guethlein *et al.* [56].

Briefly, we inserted an earthed slit at the centre of the drift region, such that only ions moving along the axis defined by the laser focus and the metal slit were detected. Behind the slit we applied a relatively weak, homogeneous magnetic field of approximately 0.26 T. The exterior of this magnetic region was shielded with an iron cast. The entire deflection unit could be positioned between 20 and 120 mm from the detector. The particle-detection unit consisted of an electron multiplier mounted on a precision translator. It was shielded all around with a metal case, except for the 1 mm × 5 mm entrance slit. A general schematic diagram of this set-up is shown in figure 5.1(a).

The deflection x of an ion with charge q, mass m and total kinetic energy T within the magnetic field B is given (for small deflection angles α) by

$$x = Bd\left(l - \frac{d}{2}\right)\sqrt{\frac{q^2}{2mT}}, \qquad (5.19)$$

where l is the distance from the detector to the entrance of the magnetic field of length d. By measuring the lateral position x of the detector, the start-up energy is directly obtained from the total electrostatic potential in the focal volume U, by using $E = T - U$. Additionally, the TOF (t) of forward and backward ions with energy E can be calculated according to Wiley and McLaren [57]. For forward ions with large start-up energies, the ratio qt/x reaches a constant value and ions with charge q will appear as a line when the signal intensities obtained are plotted on an x–t diagram.

We emphasise that the above description is valid only for small deflection angles α. In order to extend the analysis towards larger deflection angles we numerically integrated the equation of motion for the ions and thus simulated the time of flight and the deflection using 5-ps time steps.

(a)

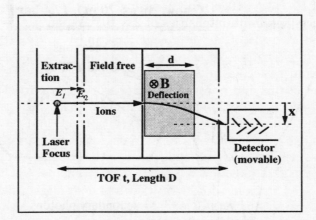

(b)

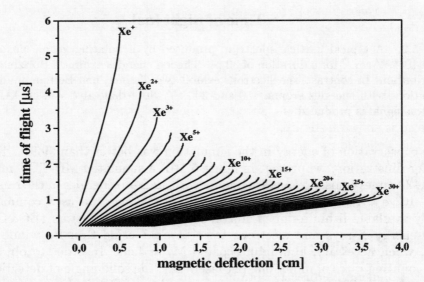

Fig. 5.1. (a) A schematic view of the magnetic deflection TOF (MD-TOF). Ions are produced in the centre of the first electric-field region $\mathcal{E}_1$ and accelerated by $\mathcal{E}_1$ and $\mathcal{E}_2$ into the field-free drift region. In the deflection unit a homogeneous magnetic field B deflects the ions with respect to kinetic energy and charge q. The detector can be positioned along the x-axis. (b) A simulated MD-TOF spectrum for the arrangement in (a) for xenon ions with charge states between 1 and 30 and energies between 1 eV and 1 MeV. The lateral position x of the detector and TOF t have been calculated using typical parameters taken from our experiment ($B = 0.26$ T, $d = 35$ mm).

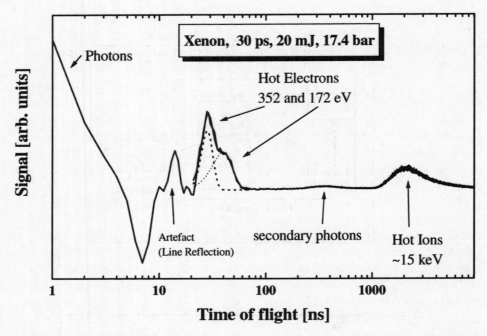

Fig. 5.2. A typical particle spectrum produced by irradiating xenon clusters with 10^{14} W cm^{-2} for a duration of 30 ps. The ion energies exhibit a Boltzmann distribution. In contrast, the electrons exhibit two distinct, non-Boltzmann distributions with energies around 150 and 300 eV. Note that, at $t = 0$, an XUV-photon signal is produced.

The conservation of energy in the simulation was better than 99.9%. For better illustration, we present in figure 5.1(b) a simulated MD-TOF map for 1–30-times-ionised xenon fragments with start-up energies in the range of 1–10^6 eV. For high-sensitivity detection of X-rays, we used commercially available lithium-doped silicon semiconductor detectors (EG&G). Low-energy X-rays and electrons were blocked using thin Be windows and, when necessary, aluminium-coated Mylar foils. In order to obtain low counting rates of under one per laser shot, the solid angle of detection was reduced with small copper diaphragms. A more detailed presentation of the technique can be found in [58].

5.4 Long-pulse responses of rare-gas clusters

We studied the interaction of rare-gas clusters with 30-ps pulses at peak intensities of 10^{14} W cm^{-2}. Figure 5.2 shows the electron, photon and ion signals after irradiation of xenon clusters with average sizes of around 2×10^6 atoms. We used the detector arrangement described in section 5.3.

Apart from the strong photon signal at $t = 0$, four additional features are observed. At flight times of around 2 μs, a broad ion signal is visible, which is attributed to Boltzmann-distributed ions with a temperature near 15 keV. A similar kinetic distribution, showing up much more weakly at around 300 ns, is produced by secondary photons that result from the recombination of highly charged ions on the surrounding metal surface located about 30 mm from the focus. Additionally, between 20 and 30 ns, fast electrons with energies between 100 and 300 eV are detected. With the help of a weak magnetic field we were able to certify that these signals were indeed produced by electrons, since they vanished completely, whereas all the other signals remained.

It is interesting to note that the kinetic-energy distribution of the electrons is evidently a non-Boltzmann shaped one. Instead it consists of two distinct, Gaussian-shaped peaks around 300 and 150 eV. This observation corresponds qualitatively to earlier results of Shao *et al.* [59], who also studied xenon clusters, but at much higher laser intensities above 10^{16} W cm^{-2}. These authors also reported finding non-Boltzmann distributions of electrons within two distinct energy ranges. Whereas the more energetic electrons are emitted isotropically and originate from hydrodynamic expansion, the authors assign the lower energy distribution to free-streaming electrons leaving the cluster along the lines of the electric field of the laser beam [59, 60]. However, in our experiment the cycle-averaged oscillatory electron energy (quiver energy)

$$U_Q = e^2 E^2 / (4 m_e \omega_L^2) = 9.33 \times 10^{-14} I \lambda^2, \qquad (5.20)$$

(I in W cm^{-2} and λ in micrometres yields U_Q in electron-volts) is only about 20 eV, which ought to be insufficient for producing significant amounts of free-streaming electrons. For this reason we are currently unable to explain the two distinct electron distributions in figure 5.2. Moreover, the electron energies are not consistent with the ion-plasma temperature derived from the observed energies of ions. Finally, it is worth noting that the double structure is observed only with xenon clusters, not with argon and krypton clusters.

We also performed a systematic study of the electron and ion energies as a function of the cluster-jet backing pressure P_0 (figure 5.3). Again, we used argon, krypton and xenon clusters and laser pulses of duration 30 ps and a peak intensity of 10^{14} W cm^{-2}. It is known that, when P_0 is raised, the overall particle density in the focus will also rise linearly, whilst the mean size of clusters will grow approximately quadratically (cf. equation (5.7)). However, neither the electron nor the ion temperatures really followed this scaling. Instead, the total kinetic energy appeared to saturate, whilst the electron–ion-energy ratio depended on the constituents of clusters. The highest ion temperatures were observed for krypton,

(a)

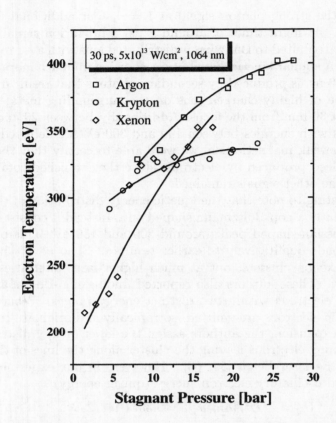

Fig. 5.3. (a) Average energies of electrons derived from experimental spectra similar to those shown in figure 5.2 for various backing pressures of the cluster gas. The parameters of the laser pulse were 30 ps, 10^{14} W cm^{-2} and 1064 nm.

in combination with relatively low electron energies. From the observed properties of the electrons and ions, we conclude that the irradiation of clusters with 30-ps pulses resists a simple interpretation, but seems to lead, in general, to the formation of a strongly heated, non-thermalised plasma.

In another long-pulse study of argon clusters, we carefully analysed the ion signal using a standard TOF technique. We will see that this allows us to determine the cluster-explosion dynamics. Figure 5.4 shows TOF spectra obtained after irradiation of argon clusters with 30-ps pulses at 10^{14} W cm^{-2}. By using various extraction voltages, we could change the forward–backward splitting of the fast, multiply charged fragments [57]:

$$\Delta t = 1.44 \times 10^3 \sqrt{m_i E_i}/(q\mathcal{E}), \qquad (5.21)$$

where E_i is the ion start-up energy in electron-volts, and $\mathcal{E}$ is the extrac-

(b)

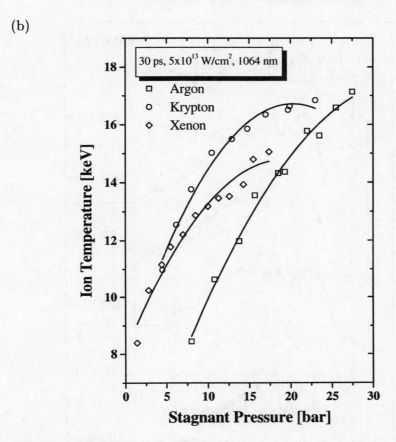

Fig. 5.3. (b) Average energies of ions observed for the same experimental situation as that in (a).

tion field at the focus in V cm^{-1}. The peak-splitting Δt is calculated in nanoseconds. A more detailed description of the technique can be found in [61].

We first note the broad ion distribution between 3 m_i/q and 10 m_i/q in figure 5.4. This feature disappears when either the cluster beam is blocked, or the spatial overlap between the pulsed cluster jet and the laser pulse is changed. It is assigned to Ar^{q+} ions, with q between 4 and 10, that are smeared out due to the very large initial kinetic energies. With decreasing extraction voltage a step structure appears on the right-hand shoulder of the broad ion distribution. These thresholds are attributed to the loss of fast backward ions after collision with the repelling plate. The centre energy of each charge state is evaluated from the field-dependent displacement of the corresponding maxima. Unfortunately, this cut-off analysis fails for the more highly charged ions, like Ar^{9+} and Ar^{10+},

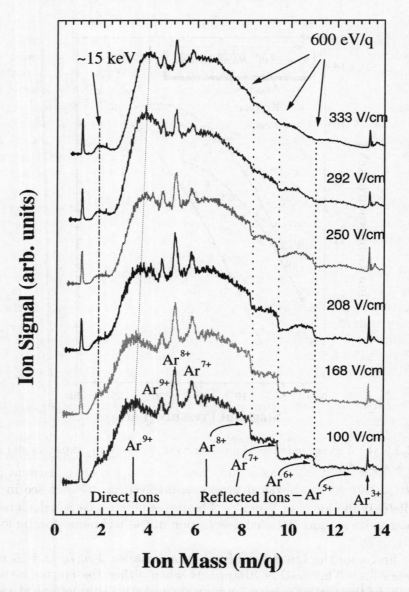

Fig. 5.4.   Ion mass spectra obtained for various extraction fields $\mathcal{E}$ using Ar with $P_0 = 8$ bars. Ar^{q+} ($q = 7$, 8 and 9) appear in distinct peaks at 5.7, 5.0 and 4.4 times m_i/q, respectively. The signal cut-offs at 13.17, 10.98, 9.41 and 8.23 times m_i/q can be attributed to Ar^{q+}, $q = 5$, 6, 7 and 8 with energies defined by $E = q\mathcal{E}l$, where $l = 3.5$ cm is the distance between the focus and the repelling plate. The feature at $2m_i/q$ is assigned to fast argon ions having more than ten positive charges (see the text).

because of their low abundance. From our analysis we find an average ion energy of $q \times 600$ eV. The linear energy dependence strongly suggests that the explosion is dominated by hydrodynamic forces [61]. This result is in agreement with our expectation that the Coulomb pressure does not play a significant role in the present, low-intensity regime. Significant amounts of ions leave the laser focus with relatively low kinetic energies and show up as distinct centre peaks for $q = 7$, 8 and 9. These peaks may be due to low-energy ions that are produced near the centre of the cluster.

Regarding the small peak at $2m_i/q$, the TOF method used at present fails to assign this channel unambiguously. Nevertheless, there are two reasons to believe that this small peak can be attributed to fast Ar^{10+} ions with kinetic energies between 8 and 25 keV. Firstly, we observe a small peak at $4m_i/q$ indicating the presence of Ar^{10+}. Secondly, spectra obtained without an extraction field show that such high-energy ions are indeed produced in these quantities (see figure 5.3). In the following section we will see that such signals are probably due to ions that undergo Coulomb-driven acceleration. This means that a small amount of free-streaming electrons must also be produced in the present, moderate-intensity regime.

In summary, the interaction of rare-gas clusters with moderately intense, picosecond laser light results in remarkable ion energies arising from the heating of the nanoplasma. However, these long picosecond laser pulses also give rise to rather complicated patterns of energy transfer, manifested by the inconsistency between the electron and ion temperatures. A detailed understanding of these processes would require complex, numerical modelling with a plasma code that can describe the physical conditions (e.g. LASDEX). Fortunately, the use of ultra-short laser pulses simplifies the situation considerably and provides a more consistent and clearer picture of the physical processes involved, as we will see in the next paragraph.

5.5 Short-pulse-regime studies with rare-gas clusters

As we mentioned earlier, the use of ultra-short laser pulses in the femtosecond regime essentially confines the nuclei during the interaction time. The ion kinetics is therefore restricted to the temporal evolution of the electron temperature and, additionally, to the total charge Q deposited in the cluster's volume by the loss of Q/e electrons. For this reason, the interaction of ultra-short and intense laser pulses with clusters has recently received considerable attention, leading to the discovery of very highly charged ions with kinetic energies up to the mega-electron-volt regime [37]. In this way, experiments on clusters have become a new laboratory for the investigation of small-scale plasma phenomena far from local thermal equilibrium and with strong temperature gradients.

Many years ago, during the early studies on laser–matter interactions, ultra-fast ions with several MeV amu^{-1} were observed. This phenomenon has been attributed to various processes such as instabilities caused by over-heated coronal electrons [62], rarefaction shocks [63], neutralising return currents [64], ion-acoustic turbulences [65] and self-focusing [66]. In contrast, the kinetic energies obtained from clusters have been interpreted from the beginning as being due either to Coulomb explosion or to hydrodynamic expansion of a super-heated nanoplasma.

While hydrodynamic expansion is a well-established mechanism for a large scale plasma, Coulomb explosion is the appropriate model for molecules in intense laser fields. It is therefore of considerable interest to study the transition region between the ranges of applicability of these two mechanisms. However, the experimental investigation of this situation is difficult when standard TOF techniques such as those described in the previous chapter are being used. This is easily understood from figure 5.5, which shows the primary ion signal from xenon clusters irradiated at approximately 10^{17} W cm^{-2}. In the first instance, the intense photon signal at $t = 0$ contaminates the electron signal so severely that simple electron-TOF analysis is no longer possible. Secondly, a considerable part of the ions arrives at around 0.7 µs, which corresponds to ion energies of around 150 keV. It is evident from this snapshot that the distribution of energies is no longer a simple Boltzmann function. Although the ion signal at around 3 µs appears to have a Boltzmann distribution with a temperature of 20 keV, the fast-ion peak at 0.7 µs is definitely not distributed in this way.

Since the start-up energies of the highly charged fragments have an exceptionally broad distribution, the traditional methods of mass spectrometry are no longer useful. In addition, the average charge state of the ions is expected to be somewhere between 15$^+$ and 20$^+$, which would demand a mass resolution of $m/\Delta m > 500$. One interesting approach to analysing a similar mix of ions was recently employed by Ditmire *et al.* [37], who introduced an electrostatic retarding potential in the TOF drift region. With a simulation of the transmitted ion signals, it is possible to estimate the kinetic-energy distributions of the various charge states. Despite the elegance of this technique, its insufficient accuracy in mapping the charge dependence of the energy distributions hinders determination of the explosion dynamics of the nanoplasma. We therefore developed a different experimental approach that allows us to measure such maps directly.

The power of our experimental technique is illustrated by the contour plot in figure 5.6, which combines numerous TOF spectra taken at different lateral positions x of the detector. The typical MD-TOF map shown

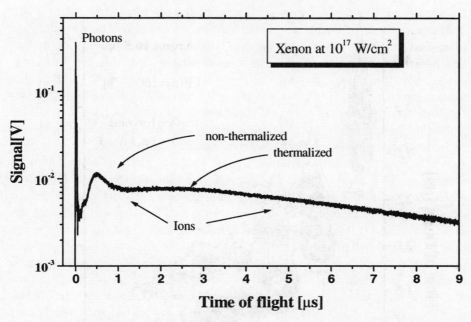

Fig. 5.5. A typical particle spectrum obtained from xenon clusters with average sizes exceeding 10^6 constituents. The aggregates were irradiated with 130-fs pulses at 10^{17} W cm^{-2} and a wavelength of 790 nm. Despite the strong signal at $t = 0$, which is produced by photons and fast electrons, the two different signals indicate that the ions are not Boltzmann-distributed. This strongly suggests that we have a completely non-thermalised situation.

in the figure was obtained after irradiation of argon clusters with 130-fs pulses at 5×10^{17} W cm^{-2}. Forward- and backward-scattered ions with $q = 1$–8 are fully resolved on the time axis and show up at distinct x values. From the peak splitting, the start-up energies of these ions can be directly evaluated and turn out to follow a quadratic charge dependence:

$$T(q) = \alpha_{\mathrm{Cb}} q^2, \quad \alpha_{\mathrm{Cb}} = 160 \text{ eV}. \tag{5.22}$$

This result clearly indicates a purely Coulomb-driven explosion. The x–t positions corresponding to equation (5.22), are indicated by triangles in figure 5.6. The magnetic field was calibrated from the energy dependence obtained from the peak splitting in combination with the lateral positions x of the detector. The overall agreement of our simulation is satisfactory.

Figure 5.6 reveals a rather interesting feature, denoted by A, which is related to the presence of Ar^{q+} ions with $q > 6$. Ion signals at A are forming tails that are directed towards the origin of the x–t diagram, indicating quite high kinetic energies. Comparison with the simulation reveals that these ions must have energies of the order of 50 keV.

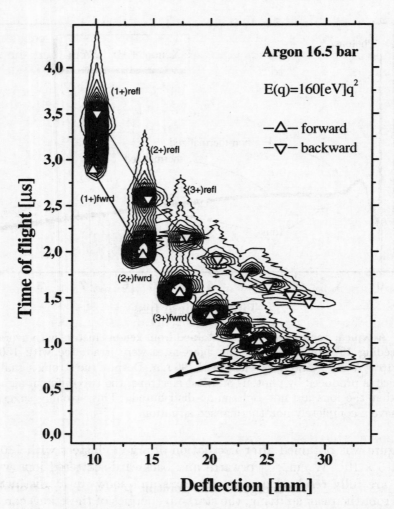

Fig. 5.6. A two-dimensional contour plot, showing a typical MD-TOF map obtained from laser irradiation of large argon clusters. Signals correspond to Ar^{q+} ions, $(q = 1\text{–}8)$. The start-up energies E for forward ($\triangle$) and backward ($\triangledown$) ions can be modelled by assuming that $E(q) = q^2 \times 160$ eV.

Both the quadratic charge dependence and the presence of feature A cannot be understood on the basis of a hydrodynamic-expansion model, wherein $E(q) \propto q$ is determined by the electron temperature. Interestingly, these clusters, although they have average sizes of around 200 000 atoms, appear to behave similarly to small molecules with respect to the purely Coulomb-driven expansion.

To study cluster-size effects, we produced larger clusters and used atoms of higher Z as cluster constituents. Our goal was to investigate the transition of the mechanism of expansion to that of solid targets. While leaving the total number of atoms in the focus approximately constant, we were able to produce clusters that were about ten times larger, simply by switching from argon to xenon as the clustering gas. The result of this experiment is shown in figure 5.7. Again, the MD-TOF map shows forward–backward-split ion signals, but only for charge states up to 3. For the higher charges, on the other hand, the backward-scattered ions are completely lost at the repelling plate. Careful analysis of the x–t positions in the signal of the forward-scattered ions reveals that their energy dependence can be understood only when two distinct mechanisms of acceleration are included.

Ions with $q < 6$ seem to follow a quadratic charge dependence, similarly to the argon case, but with $\alpha_{\mathrm{Cb}} = 180$ eV. We note that particularly ions with $q = 1$–3 appear to be delayed in time (by up to 800 ns) with respect to the simulated x–t values. We further observed that this retardation depends on the particle density in the focus and we suggest therefore that it is due to a temporary shielding of the TOF extraction field, due to the local polarisability of the plasma. Only the flight times of low-charged ions are influenced by this effect, probably because they leave the focus with the smallest momentum. However, since these ions are produced in the outer regions of the focus at low laser intensities, the strong-field outcome of the cluster experiment is not significantly affected.

Interestingly, the energy of most Xe^{q+} ions with $q > 10$ follows a linear energy dependence with $T(q) = q \times 1500$ eV. As already stated in section 5.4, this dependence is characteristic of a hydrodynamic expansion. Thus, xenon clusters with an average size of around 2×10^6 indeed clearly exhibit solid-like behaviour with respect to their mechanism of explosion. Nevertheless, a minor part of the highly charged ions with $20 \leq q \leq 30$ again shows up in highly energetic tails pointing towards the x–t origin. For comparison, we have introduced an energy scale in figure 5.7 for Xe^{30+} ions of energies ranging from 50 keV to 1 MeV. Clearly, some of the highly charged ions have start-up energies between 400 keV and 1 MeV. These very high energies are in fact quite consistent with the expected Coulomb pressure inside large-scale clusters, if we assume that there is a significant build-up of positive charge.

Extracting the energy distributions of the various charge states from figure 5.7 is straightforward. This important information is directly accessible if we take the simulated set of $t(E, q)$ and $x(E, q)$ values, see figure 5.1(b) and apply them to figure 5.7. Furthermore, this procedure also automatically yields the average signal intensities. The obtained kinetic-energy distribution of fragments as a function of the charge state is depicted in figure 5.8. Again, the high-energy tail, corresponding to Coulomb explosion, is clearly visible for the more highly charged frag-

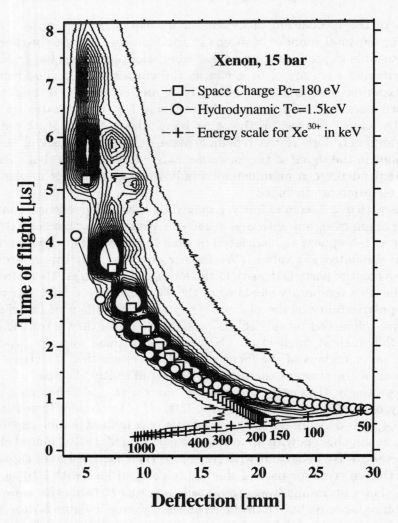

Fig. 5.7.   The MD-TOF spectrum obtained from laser irradiation of large xenon clusters. Most signals correspond to forward Xe^{q+} ions ($q = 1$–30); only weak backward signals are visible for $q = 1, 2$ and 3. The energies of ions are simulated using both a hydrodynamic approach ($\circ$: $E(q) = \alpha_H q$, $\alpha_H \approx 1.5$ keV) and a charged-sphere model ($\square$: $E(q) = \alpha_{Cb} q^2$, $\alpha_{Cb} \approx 180$ eV). The inserted energy scale ($-+-$) indicates the position of Xe^{30+} ions with start-up energies between 50 keV and 1 MeV.

ments with energies of up to several times 100 keV. The ability of MD-TOF spectrometry to resolve even such broadened start-up energy distributions for each charge state illustrates the potential of this method for the analysis of fast and multiply charged ions from intense laser–matter

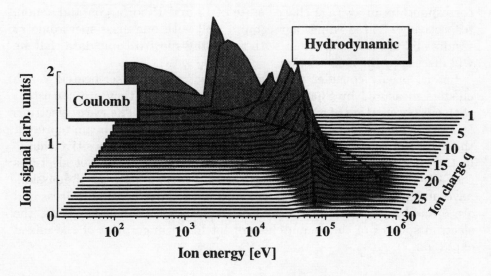

Fig. 5.8. Energy distributions of xenon charge states ($q = 1$–30) deduced from the MD-TOF spectrum presented in figure 5.7. The analysis is based on the simulated x–t positions presented in figure 5.1(b). It can be seen that the kinetic-energy distribution of the ions exhibits a hydrodynamic and a Coulomb-driven mechanism of acceleration.

interactions.

It is also interesting to note that the argon and xenon clusters exhibit very similar Coulomb-energy dependences ($q \times 160$ eV versus $q \times 180$ eV), even though the average sizes of cluster are quite different. This is possible only if the charge density created inside a cluster is virtually independent of its size and constituents. In other words, the Coulomb pressure inside a cluster, see equation (5.18), depends predominantly on the electric-field strength of the laser beam and hardly at all on the cluster species itself.

More evidence for this behaviour is found directly from MD-TOF maps for clusters of argon (figure 5.6) and xenon (figure 5.7). It is well known that clusters that are produced by supersonic expansion usually exhibit rather broad size distributions. In comparison, we find from the MD-TOF analysis quite well-defined Coulomb energies. On the one hand, this independence of the cluster radius makes it impossible to evaluate the size of a cluster from the observed Coulomb energies. On the other hand, it allows us to estimate the average charge state of the ions on the charged surface of a cluster, if we assume that the total charge Q resides on the surface. By inserting into equation (5.18) the typical van der Waals radius as the average distance between adjacent surface atoms, we obtain a local Coulomb pressure of $\simeq 10^{11}$ J m^{-3} per van der Waals sphere. This would correspond to an average charge state of 13 and 15 for argon and xenon, respectively. These values agree quite well with our mass-spectrometry results and are also consistent with the X-ray-spectroscopic data that we will discuss in the next section.

Let us briefly consider the electron spectra from laser-heated xenon clusters measured by Shao *et al.* [59]. We find that they are consistent with our observation of a mixed expansion behaviour. The warm electrons (≤ 1 keV) emitted along the electric-field vector of the laser beam represent the lost electrons that are responsible for the build-up of positive charge in the cluster on the leading edge of the laser pulse. The hot electrons ($\simeq 2.3$ keV), which are characterised by an isotropic pattern of emission, have undergone numerous collisions during the laser-heating process and are ejected as a result of the hydrodynamic expansion [60]. Thus, the electron spectra of the xenon clusters act like a fingerprint of the mixed explosion.

5.6 The generation of X-rays with clusters in strong fields

A drawback of measuring electrons or ions is the fact that detection occurs late after the ultra-short interaction with the laser beam. Thus, the results discussed in the previous sections do not necessarily reflect the situation in the nanoplasma at the time of the interaction. X-ray spectrometry should be used to obtain more reliable data and to get almost instantaneous information on parameters such as the average charge state of the ions, the electronic temperature and the density of ions in the nanoplasma. As an example, due to fast recombination during the explosion, the average charge state can differ from that deduced by mass spectrometry.

Since the pioneering studies of Rhodes and co-workers on generation of X-rays with clusters in the gas phase, it has become clear that high-density jets of noble-gas clusters are very promising targets for applications for several reasons. Unusually high X-ray-production efficiencies were reported for jets of clusters [32]. Gas jets also go well with high-repetition-rate X-ray schemes, they are debris-free and the emission times are possibly in the same range as the duration of the ultra-short laser pulse. We have therefore studied the emission patterns of cluster-produced X-rays in order to obtain quantitative information on the conversion efficiency.

Whereas gratings are often used for photons in the energy range up to 500 eV, higher photon energies require other spectroscopic techniques. For high-resolution spectrometry in the range up to several kilo-electron-volts, diffraction on crystals can be employed. However, such spectrometers are characterised by a very low transmission and the useful range for a given crystal is quite restricted. In comparison, semiconductor detectors have relatively low resolution but, on the other hand, a high detection effi-

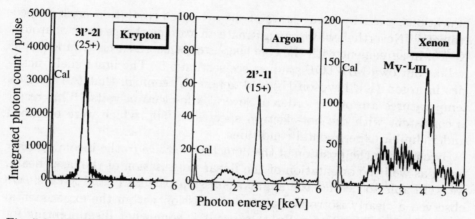

Fig. 5.9. Typical X-ray spectra obtained from laser-irradiated rare-gas clusters. Spectra were obtained using a Si(Li) semiconductor detector. The measurement was performed at a reduced counting rate of 0.05 s^{-1}. As indicated, the spectral lines can be attributed to K-shell transitions of argon and to L-shell transitions of krypton and xenon.

ciency approaching unity. Semiconductor detectors, therefore, seem to be the best choice for measuring absolute X-ray-photon yields and emission patterns. Although we applied both spectroscopic techniques in our experiments [67], we will concentrate on the quantitative results obtained with Si(Li) semiconductor detectors with well-known characteristics.

Two detectors were mounted at different positions relative to each other and with respect to the incident laser beam. The spectral resolution is typically better than 10% of the transition energy, provided that the photon-counting rate is much lower than one per laser pulse. Otherwise pile-up occurs and the spectral intensities have to be evaluated using Poissonian statistics (see for instance [58]). Figure 5.9 depicts X-ray spectra obtained for argon, krypton and xenon clusters. All three spectra have large photon signals above 1 keV, which are produced by inner-shell transitions of highly charged ions. For example, the peak in the argon X-ray spectrum can clearly be attributed to $2\ell \rightarrow 1\ell'$ transitions with average charge states of the order of 14^+ to 15^+. These values correspond well to the ion-charge states reported in the previous section and are also confirmed by our high-resolution studies [67]. With krypton we observed X-rays that we attribute to $3\ell \rightarrow 2\ell'$ transitions in ions with an average charge of around 25^+. The highest photon energies, exceeding 4 keV, were observed with xenon clusters. These very energetic photons originate from inner-shell transitions between the M and L shells.

The presence of many residual electrons in krypton and xenon makes it very difficult, or even impossible, to identify the numerous electronic configurations, even by using state-of-the-art, high-resolution crystal spectrometry. Nevertheless, we can estimate an average charge state of around Xe^{30+}, in agreement with the conclusions of other research groups [32] as well as our own MD-TOF studies, see section 5.5. The inner-shell vacancies in argon (K), krypton (L) and xenon (L) confirm that the electron temperatures are of the order of several kilo-electron-volts. This result is consistent with the hot-electron spectra of [59], which were obtained under similar experimental conditions.

By changing the position of the detectors relative to the incoming laser beam as well the orientation of the linear polarisation of the laser beam, we were able to deduce the angular-emission pattern of the X-rays. We observed a clearly isotropic pattern of emission within the experimental error of ±5% [58]. Naturally, this result is somewhat disappointing for applications, for which confined X-ray emission would be very advantageous. The isotropic pattern of emission is, however, not surprising when we consider the conclusions of the previous section. Even for highly oriented systems, the coupling of angular momentum in many-electron ions leads to severe depolarisation during the emission. Moreover, since ionisation is a result of many electron–ion collisions during the short laser pulse, the orientational information of the incident laser beam is easily lost.

For the determination of the total X-ray yield, an isotropic pattern of emission is quite helpful since a detector with a restricted solid angle of detection will suffice. We thus determined the efficiency η of conversion of power from the incoming laser beam to the emission of X-rays, integrated over 4π. For argon K-shell transitions, we obtained $\eta_{Ar} = 2.2 \times 10^{-11}$, a value that is roughly 35 times smaller than that for xenon L-shell transitions, $\eta_{Xe} = 7.7 \times 10^{-10}$. The most efficient conversion was observed for krypton L-shell transitions, $\eta_{Kr} = 2.1 \times 10^{-8}$. This value is about 1000 times larger than that for argon. Nevertheless, all these values remain extremely small relative to typical conversion efficiencies for solids, which occasionally exceed 10%. However, we would like to emphasise that the overall density of our skimmed cluster jet of approximately 10^{14} cm^{-3} was relatively low. It is straightforward to increase the density of the target by a factor of 10^5, thereby almost approaching the densities of liquids. Preliminary results [67] indeed show that much higher X-ray yields are achievable in that way. Then, the photon flux should be sufficient to measure the temporal emission profile by means of ultra-fast X-ray streak cameras.

We investigated the mean number of X-ray photons per laser shot $\bar{N}_p$, as a function of the peak intensity of the laser beam I_L. The value $\bar{N}_p$ is derived from pile-up spectra using Poissonian statistics [58]. The resulting

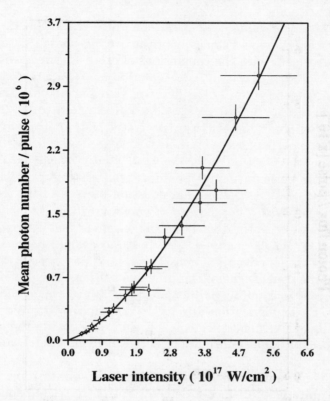

Fig. 5.10.   The yield of X-ray photons as a function of the peak intensity of the laser beam, for clusters of estimated mean size 7×10^5. The solid line is a fit of the experimental data, using $\bar{N}_\mathrm{p} \propto I_\mathrm{L}^{3/2}$. Such a power law indicates that saturation of the X-ray-emission process occurs (see the text).

dependence is shown in figure 5.10. It can be fitted by

$$\bar{N}_\mathrm{p} \propto I_\mathrm{L}^{3/2}. \tag{5.23}$$

This scaling law is typical for a saturated process in the focus of an intense laser field with a Gaussian intensity profile. This scaling was first observed for OFI of rare-gas atoms [68], for which the further increase of the yield of ions after saturation is due to the increase of the focal volume with increasing I_L, see chapter 2.

Interestingly, we find that the efficiency of production of X-rays is proportional to the square of the size of the cluster. We conclude this from the observation that, in our experiments (see figure 5.11),

$$\bar{N}_\mathrm{p} \propto (P_0 - P_\mathrm{c})^3, \tag{5.24}$$

with $P_\mathrm{c} \approx 3$ bar. From the scaling laws in section 5.2 we know that the

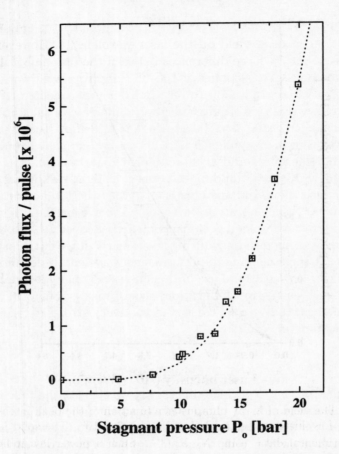

Fig. 5.11.   The yields of X-ray photons $\bar{N}_\mathrm{p}$ as a function of the backing pressure P_0, obtained at 4×10^{17} W cm^{-2}. The dashed line indicates a fit using $\bar{N}_\mathrm{p} \propto (P_0 - P_\mathrm{c})^3$. Such a pressure dependence can in principle be understood by invoking a mechanism for deposition of energy based on collisional absorption (see the text).

size of a cluster depends roughly on P_0^2 and that the total number of clusters per unit volume of the molecular beam is proportional to P_0^{-1}. Thus equation (5.24) can be understood only when each cluster radiates like N^2. The reason for this behaviour is again found by considering the mechanism of inverse *Bremsstrahlung*. Initially, each cluster is ionised by the interaction with the electro-magnetic field, with a cross section that is essentially given by the size of the cluster. Subsequently, the cluster's constituents are further ionised by electron–ion collisions. The cross section for the second process must therefore also be linearly proportional to the size of the cluster.

Finally, it should be mentioned that there is possibly an important dependence of the photon yield on the laser wavelength. However, in the experiments at Saclay, large differences between the fundamental and the second harmonic ($\lambda = 395$ nm) of the Ti:sapphire laser were not observed. This is in contradiction to results from experiments of Kondo *et al.* [69], who have reported a quite substantial increase in the photon yield when they changed from $\lambda = 800$ nm to $\lambda = 248$ nm. Today, the issue of such a wavelength dependence appears to be one that should be investigated more carefully and at more intermediate wavelengths. In fact, from equation (5.13) one would indeed expect a slightly different coupling between the laser field and the cluster at either the fundamental or the second harmonic wavelength. However, at UV wavelengths other mechanisms such as, for example, mechanisms involving collective resonances of plasmons might come into play during a very early stage of the expansion. Nevertheless, it is challenging to perform such wavelength-dependence experiments accurately. Despite use of the cluster jet, many key parameters of the laser such as the pulse duration, size of focal spot, focal position, chirp and intensity have to be conserved to a very strong degree in order to obtain meaningful results.

5.7 Conclusion

In view of the present experimental results, we suggest the following scenario for the interaction of clusters with an intense laser pulse. Even during an early stage of the laser pulse most of the atoms are ionised by field ionisation. However, only a limited number of electrons can leave the system after ionisation. The remaining electrons are trapped as a Fermi droplet by the Coulomb potential of the cluster.

With every optical cycle these electrons are further heated by collisional absorption, which results in a rise of the electron temperature as well as a strong increase in electron density. This finally leads to efficient, multiple-collision-induced, ionisation. The remaining quasi-free electrons of the nanoplasma, as well as the excited core electrons, undergo inner-shell transitions and give rise to generation of kilo-electron-volt X-rays. Some of these X-rays are possibly absorbed by surrounding atoms, which could lead to further efficient heating. A high energy density is reached before the cluster really starts to expand. During the expansion the density of the nanoplasma approaches the critical density, at which the absorption of light becomes resonant. At this point, the deposition of energy increases again dramatically within a few optical cycles. This induces a massive loss of electrons, which can lift the Coulomb pressure on the surface of the nanoplasma well above the hydrodynamic pressure.

Brewczyk *et al.* [70] recently applied a Fermi-liquid model to small one-dimensional systems. Their calculations yield highly charged ions that undergo Coulomb ejection in a shell-wise manner within a few optical cycles. This suggests that clusters explode in two steps. Initially, the most highly charged ions, which are located close to the surface, undergo violent ejection in an electrostatic shock wave, which is also observed in experiments with solid targets. Later, when the positive excess charge of the system has dropped sufficiently, the hydrodynamic forces take over and drive the expansion in such a way that the system behaves like a small, spherical plasma. According to this mechanism, the relative velocities of neighbouring atoms will be very similar and, since the acceleration is purely radial, intra-cluster collisions will occur rather seldomly. The particle energies will, therefore, not be in perfect thermal equilibrium, which is consistent with our observation of an energy splitting into a Coulombic and a hydrodynamic part.

In conclusion, clusters in strong laser fields present a new, interesting laboratory for studying intense-laser–matter interaction in the gas phase. Many questions remain unanswered and demand further investigation. Metal clusters that consist of high-Z atoms are expected to be very efficient at emitting high-energy X-rays and to explode with large ion energies. Finally, we are looking forward to experiments with clusters in the relativistic regime well above 10^{18} W cm^{-2}, which will undoubtedly reveal further interesting effects.

Acknowledgements

M. L. thanks the European Union for the support he received under grant number ERBF MBICT 95042. It is a real pleasure to thank S. Dobosz, J.-P. Rozet, D. Vernhet, M. Perdrix, P. Meynadier and D. Normand for their important contributions regarding many of the experiments and discussions. The authors would also like to thank J. H. Posthumus for his editorial work.

References

[1] G. Mainfray and C. Manus, *Rep. Prog. Phys.* **54** 1333 (1991) and references therein.

[2] M. M. Murnane, H. C. Kapteyn, M. D. Rosen and R. W. Falcone, *Science* **251** 531 (1991); H. M. Milchberg, R. R. Freeman, S. C. Davey and R. M. More, *Phys. Rev. Lett.* **67** 2654 (1991) and references therein.

[3] A. L'Huillier, L. A. Lompré, G. Mainfray and C. Manus, *Phys. Rev. Lett.* **48** 1814 (1982).

[4] T. S. Luk, H. Pummer, K. Boyer, M. Shahidi, H. Egger and C. K. Rhodes, *Phys. Rev. Lett.* **51** 110 (1983).

[5] M. Perry, A. Szöke, O. Landen and E. Campbell *Phys. Rev. Lett.* **60** 1270 (1988).

[6] S. Augst, D. Strickland, D. D. Meyerhofer, S. L. Chin and J. H. Eberly, *Phys. Rev. Lett.* **63** 2212 (1989).

[7] L. V. Keldysh, *Sov. Phys. JETP* **20** 1307 (1965).

[8] F. Faisal, *J. Phys. B* **6** L312 (1973).

[9] H. R. Reiss, *Phys. Rev. A* **22** 1786 (1980).

[10] M. V. Ammosov, N. B. Delone and V. P. Krainov, *Sov. Phys. JETP* **64** 1191 (1986); *J. Opt. Soc. Am. B* **8** 1207 (1991); *Sov. Phys. JETP* **74** 789 (1992).

[11] A. Kazakov and M. Fedorov, *Prog. Quantum Electron.* **13** 1 (1989).

[12] P. Agostini, F. Fabre, G. Mainfray, G. Petite and N. Rahman, *Phys. Rev. Lett.* **42** 1127 (1979).

[13] Y. Gontier, M. Poirier and M. Trahin, *J. Phys. B* **13** 1381 (1980).

[14] A. McPherson, G. Gibson, H. Jarah, U. Johann, T. S. Luk, I. McIntyre, K. Boyer and C. Rhodes, *J. Opt. Soc. Am. B* **4** 595 (1987).

[15] M. Ferray, A. L'Huillier, X. F. Li, L. A. Lompré, G. Mainfray and C. Manus, *J. Phys. B* **21** L31 (1988).

[16] A. L'Huillier and Ph. Balcou, *Phys. Rev. Lett.* **70** 774 (1993); J. J. Macklin, J. D. Kmetec, C. L. Gordon, *Phys. Rev. Lett.* **70** 766 (1993).

[17] P. B. Corkum, *Phys. Rev. Lett.* **71** 1995 (1993).

[18] P. Salieres, T. Ditmire, M. D. Perry, A. L'Huillier and M. Lewenstein, *J. Phys. B* **29** 4771 (1996).

[19] P. Salieres, A. L'Huillier, P. Antoine and M. Lewenstein, *Adv. Atom. Mol. Opt. Phys.* **41** 84 (1998) and references therein.

[20] L. J. Frasinski, K. Codling, P. Hatherly, J. Barr, I. N. Ross and W. T. Toner, *Phys. Rev. Lett.* **58** 2424 (1987).

[21] A. Giusti-Suzor, X. He, O. Atabek and F. Mies, *Phys. Rev. Lett.* **64** 515 (1990).

[22] C. Cornaggia, J. Lavancier, D. Normand, J. Morellec and H. X. Liu, *Phys. Rev. A* **42** 5464 (1990).

[23] C. Cornaggia, D. Normand and J. Morellec, *J. Phys. B* **25** L415 (1992).

[24] A. Giusti-Suzor, F. H. Mies, L. F. DiMauro, E. Charron and B. Yang, *J. Phys. B* **28** 309 (1995).

[25] M. Schmidt, D. Normand and C. Cornaggia, *Phys. Rev. A* **50** 5037 (1994).

[26] T. Seideman, M. Yu. Ivanov and P. B. Corkum, *Phys. Rev. Lett.* **75** 2819 (1995).

[27] J. H. Posthumus, L. J. Frasinski, A. J. Giles, and K. Codling, *J. Phys. B* **28** L349 (1995).

[28] W. L. Kruer, *The Physics of Laser Plasma Interactions*, Addison-Wesley, New York, 1988.

[29] A. Rousse, P. Audebert, J. P. Geindre, F. Falliès, J. C. Gauthier, A. Mysyrowicz, G. Grillon and A. Antonetti, *Phys. Rev. E* **50** 2200 (1994).

[30] S. J. Gitomer, R. D. Jones, F. Begay, A. W. Ehler, J. F. Kephart and R. Kristal, *Phys. Fluids* **29** 2679 (1986).

[31] A. W. Castleman and R. G. Keesee, *Science* **241** 36 (1988).

[32] B. D. Thompson, A. McPherson, K. Boyer and C. K. Rhodes, *J. Phys. B* **27** 4391 (1994); A. B. Borisov, A. McPherson, B. D. Thompson, K. Boyer and C. K. Rhodes, *J. Phys. B* **28** 2143 (1995); A. B. Borisov, J. W. Longworth, A. McPherson, K. Boyer and C. K. Rhodes, *J. Phys. B* **29** 247 (1996); A. B. Borisov, A. McPherson, K. Boyer and C. K. Rhodes, *J. Phys. B* **29** L113 (1996).

[33] C. Wulker, W. Theobald, D. Ouw, F. P. Schafer and B. N. Chichkov, *Opt. Commun.* **112** 21 (1994).

[34] M. M. Murnane, H. C. Kapteyn, S. P. Gordon, J. Bokor, E. N. Glytsis and R. W. Falcone, *Appl. Phys. Lett.* **62** 1068 (1993); S. P. Gordon, T. Donnelly, A. Sullivan, H. Hamster and R. W. Falcone, *Opt. Lett.* **19** 484 (1994).

[35] J. Purnell, E. M. Snyder, S. Wei and A. W. Castleman Jr, *Chem. Phys. Lett.* **229** 333 (1994).

[36] E. M. Snyder, S. A. Buzza and A. W. Castleman Jr, *Phys. Rev. Lett.* **77** 3347 (1996).

[37] T. Ditmire, J. W. G. Tisch, E. Springate, M. B. Mason, N. Hay, R. A. Smith, J. Marangos, M. H. R. Hutchinson, *Nature* **386** 54 (1997); T. Ditmire, J. W. G. Tisch, E. Springate, M. B. Mason, N. Hay, J. Marangos, M. H. R. Hutchinson, *Phys. Rev. Lett.* **78** 2732 (1997).

[38] T. Ditmire, T. Donnelly, A. M. Rubenchik, R. W. Falcone, M. D. Perry, *Phys. Rev. A* **53** 3379 (1996).

[39] C. Rose-Petruck, K. J. Schafer, K. R. Wilson and C. P. J. Barty, *Phys. Rev. A* **55** 56 (1997).

[40] E. W. Becker, K. Bier and W. Henkes, *Z. Phys.* **146** 333 (1956).

[41] H. W. Kroto, J. R. Heath, S. C. O'Brie, R. F. Curl and R. E. Smalley, *Nature* **318** 162 (1985).

[42] A. W. Castleman Jr and K. H. Bowen Jr, *J. Phys. Chem.* **100** 12911 (1996).

[43] T. D. Märk and A. W. Castleman Jr, *Adv. Atom. Mol. Phys.* **20** 65 (1983); T. D. Märk, *Int. J. Mass Spectrom. Ion Processes* **79** 1 (1987).

[44] O. F. Hagena, *Z. Phys. D* **84** 291 (1987).

[45] O. F. Hagena, *J. Chem. Phys.* **56** 1793 (1972); *Surf. Sci.* **106** 101 (1981); *Z. Phys. D* **84** 291 (1987); *Rev. Sci. Instrum.* **63** 2374 (1992).

[46] O. F. Hagena, *J. Vac. Sci. Technol. A* **12** 282 (1994).

[47] R. Klingelhöfer and H. Röhl, *Z. Naturforsch.* **25a**, 402 (1970).

[48] D. Gerlich *et al.*, private communication, unpublished (1997).

[49] V. P. Silin, *Sov. Phys. JETP* **20** 1510 (1965).

[50] J. M. Dawson, *Phys. Fluids* **7** 981 (1964); T. W. Johnston and J. M. Dawson, *Phys. Fluids* **16** 722 (1973).

[51] N. W. Ashcroft and N. D. Mermin, *Solid State Physics*, Saunders College, Philadelphia, 1981.

[52] T. Auguste, unpublished (1997).

[53] A. F. Haught and D. H. Polk, *Phys. Fluids* **13** 2825 (1970).

[54] E. Springate, N. Hay, J. W. G. Tisch, M. B. Mason, T. Ditmire, J. P. Marangos, M. H. R. Hutchinson, *Phys. Rev. A* **61** 44101 (2000); J. Zweiback, T. Ditmire, M. D. Perry, *Phys.Rev. A* **59** R3166 (1999); see also *Proceedings of the International Quantum Electronics Conference, IQEC'98.* ISBN 1557525412, Technical Digest Series Vol. 7, Optical Society of America, 1998.

[55] M. Lezius, S. Dobosz, D. Normand and M. Schmidt, *Phys. Rev. Lett.* **80** 261 (1998).

[56] G. Guethlein, J. Bonlie, D. Price, R. Shepherd, B. Young and R. Stewart, *Rev. Sci. Instrum.* **66** 333 (1995).

[57] W. C. Wiley and I. H. McLaren, *Rev. Sci. Instrum.* **26** 1150 (1955).

[58] S. Dobosz, M. Lezius, M. Schmidt, P. Meynadier, M. Perdrix, D. Normand, J. P. Rozet and D. Vernhet, *Phys. Rev. A* **56** R2526 (1997).

[59] Y. L. Shao, T. Ditmire, J. W. G. Tisch, E. Springate, J. P. Marangos, M. H. R. Hutchinson, *Phys. Rev. Lett.* **77** 3343 (1996).

[60] T. Ditmire, E. Springate, J. W. G. Tisch, Y. L. Shao, M. B. Mason, N. Hay, J. P. Marangos and M. H. R. Hutchinson, *Phys. Rev. A.* **57** 369 (1998).

[61] M. Lezius, S. Dobosz, D. Normand, M. Schmidt, *J. Phys. B***30** L251 (1997).

[62] J. S. Pearlman and R. L. Morse, *Phys. Rev. Lett.* **25** 1652 (1978).

[63] B. Bezzerides, D. W. Forslund and E. L. Lindman, *Phys. Fluids* **21** 2179 (1978).

[64] E. Valeo and I. Bernstein, *Phys. Fluids* **19** 1348 (1976).

[65] P. M. Campbell, R. R. Johnson, F. J. Mayer, L. V. Powers and D. C. Slater, *Phys. Rev. Lett.* **39** 274 (1977).

[66] H. Hora, *J. Opt. Soc. Am.* **65** 882 (1975).

[67] S. Dobosz, M. Lezius, M. Schmidt, P. Meynadier, M. Perdrix, J. P. Rozet and D. Vernhet, *Phys. Rev. A* **56** R2526 (1997); S. Dobosz, PhD. Thesis, Université de Paris XIII, Paris, 1998, unpublished.

[68] T. Auguste, P. Monot, L. A. Lompre, G. Mainfray and C. Manus, *J. Phys. B* **25** 4181 (1992).

[69] K. Kondo, A. B. Borisov, C. Jordan, A. McPherson, W. A. Schroeder, K. Boyer and C. K. Rhodes, *J. Phys. B* **30** 2707 (1997); K. Kondo, H. Hondo, E. Miura, K. Katsura and E. Takahashi, in *Proceedings of Applications of High Field and Short Wavelength Sources VIII*, OSA, Potsdam, 1999.

[70] M. Brewczyk, C. Clark, M. Lewenstein and K. Rzążewski, *Phys. Rev. Lett.* **80** 1857 (1998).

6

Single-cluster explosions and high-harmonic generation in clusters

John W. G. Tisch and Emma Springate[†]

Blackett Laboratory Laser Consortium
Imperial College, London SW7 2BW, UK

[†] *Present address: FOM-Institute for Atomic and Molecular Physics*
Kruislaan 407, 1098 SJ Amsterdam, The Netherlands

6.1 Introduction

Atomic clusters have provoked great interest since their first observation in the mid-1950s [1]. Physicists' and chemists' fascination with them derives from the unique position clusters hold as an intermediate state between molecules and solids. Many studies have been concerned with the optical properties of clusters. An important finding was the discovery of collective electron dynamics in clusters [2, 3], which is virtually absent from laser–atom interactions. These are responsible for the 'giant resonance' seen in absorption spectra of clusters [2] and can lead to remarkable optical properties.

During the last five or so years, the study of laser–cluster interactions has been extended to laser intensities in excess of 10^{15} W cm^{-2} [4–22] (laser pulse widths in the range 0.1–10 ps) in the so-called 'strong-field' interaction regime, for which the electric field of the laser is no longer small relative to the atomic field and the interaction becomes highly non-perturbative [23]. This regime, which was made widely accessible by the development of chirped-pulse-amplification (CPA) lasers [24], had been studied for atoms [25–27], small molecules [28, 29] and bulk solids [30] since the late 1980s. In stark contrast to earlier studies of laser–cluster interactions at lower intensities that had revealed dynamics similar to those seen in molecules – with relatively inefficient coupling of laser energy to electrons and ions – studies on the generation of X-rays from gases of clusters (>1000 atoms) [4–6, 8] at $\simeq 10^{16}$ W cm^{-2} began to reveal startling evidence of a laser–cluster interaction that was very much more energetic. In fact, it was more reminiscent of the interaction with a bulk solid (i.e. displaying plasma-like behaviour), despite the clusters in question only being a few nanometres in diameter – much smaller indeed than the laser wavelength.

In the face of this new evidence, a plasma model of the laser–cluster interaction was developed [9], in which the cluster is treated as a spherical 'nanoplasma', subject to the standard processes of a laser-heated plasma, such as collisional heating and collisional and tunnel ionisation, but also taking into account the modification of the laser field inside the cluster owing to collective electron oscillations in the sub-wavelength-scale structure (the Mie resonance). The main insight that emerged from this model – explaining the X-ray observations – was that, at the Mie resonance, the collisional heating in a single cluster was greatly enhanced, resulting in plasma electron temperatures that exceeded even those obtained in laser–solid interactions. The model showed that the hot electrons then drove an extremely rapid and energetic explosion of the nanoplasma, creating a shrapnel of fast electrons and highly charged, fast ions. These predictions were subsequently confirmed by a variety of experiments [10–17, 20, 21]. It has thus become clear that clusters cannot simply be viewed as a bridge between molecules and solids. They have unique properties that allow the coupling of laser energy into matter with an unprecedentedly high efficiency.

This chapter has two aims. The first is to describe the physics of single clusters ($\geq$200 atoms) interacting with super-intense, femtosecond laser pulses, chiefly within the framework of the nanoplasma model of the laser-heated cluster, but other models are also discussed. By basing the discussion on an almost chronological review of the key experimental studies in this area, we hope to chart the rapid development of this field and also to provide the reader with an experimentalist's view of the fascinating research that has been done. The exciting applications of laser–cluster physics, such as bright, debris-free generation of X-rays and pulsed production of neutrons from thermonuclear fusion that takes place in *extended* cluster media (see chapter 7), all stem from the incredibly efficient coupling of laser energy into *single* clusters. These applications are thus underpinned by the understanding of the response of single clusters to intense laser pulses that has been gained, which is the main topic of this chapter.

The second aim of this chapter is to bring to the attention of the reader a relatively little-studied topic, namely *coherent* production of soft X-rays in cluster gases through the process of high-harmonic generation (HHG) with intense laser pulses. It is only recently that cluster media have come under the spotlight as a non-linear medium for HHG. The early indications are extremely promising – the unique properties of cluster gases may permit the extremely efficient generation of high-order-harmonic radiation.

This chapter is organised in the following way. Section 6.2 is concerned with the energetic response of isolated clusters to intense femtosecond laser pulses. Models of the interaction are described with the focus being on the nanoplasma model. The experimental apparatus and techniques used to study single-cluster explosions are also described. Measurements of the energetic electrons and ions produced in cluster explosions are reviewed in sections 6.3 and 6.4. In section 6.5, investigations aimed at boosting the energy of cluster explosions by using pulse sequences and by tailoring the composition of the cluster are described. Throughout, where possible, the experimental data are compared with the predictions of the nanoplasma and other models of the laser–cluster interaction, to draw out their strengths and weaknesses and to highlight areas for future theoretical work. Section 6.6 deals with HHG in extended cluster media, showing that cluster gases potentially offer many advantages over atomic gases. In section 6.7, some future directions of research in this exciting field are outlined.

6.2 Explosions of isolated clusters

Modelling

Overview. Authors of many studies of the dissociation of small clusters (typically 2–60 atoms or molecules) have interpreted the resulting kinetic energies of ions in terms of Coulomb explosion of the cluster [7, 31, 32]. In a Coulomb explosion, the electrons are assumed to leave the vicinity of the ion immediately after ionisation, so they have no effect on the subsequent ion dynamics. However, classical particle-dynamics simulations of small exploding clusters in which trajectories both of ions and of electrons are tracked [33] showed that, for clusters as small as 55 atoms, a large fraction of the ionised electrons remained in the cluster after ionisation due to the strong positive charge on the cluster. This indicates that the Coulomb-explosion model is satisfactory only for very small clusters, for which the electrons leave the volume of the cluster rapidly. For larger clusters, collisional ionisation by laser-driven electrons is important and higher charge states will be reached in a cluster than can be obtained through over-the-barrier ionisation of a monatomic gas irradiated at the same intensity.

Explosions of clusters in an intense laser field have also been modelled theoretically by Rose-Petruck *et al.* [34] and by Hu and Xu [35, 36]. These simulations showed that the Coulomb potential binding an electron to an ion can be lowered by the Coulomb attraction of neighbouring ions, thereby enhancing the rate of over-the-barrier ionisation. When the laser intensity exceeds the threshold for saturation of single ionisation, an avalanche of ionisation is initiated – this is known as 'ionisation ignition'. Ionisation ignition can result in a much higher charge state being attained in clusters than can occur in individual atoms.

Studies of clusters in strong laser fields were triggered by the group of Rhodes *et al.* at the University of Illinois in Chicago, who realised that observations of seemingly anomalous emission of X-rays from gas targets had been due to the presence of clusters [4]. Line emission from charge states higher than would be expected for the irradiation of atoms at the same intensity was observed as well as X-rays identified as resulting from inner-shell transitions. A numerical model was developed to explain these observations and define allowed regions of cluster size and laser intensity for which inner-shell radiation can be emitted [4, 6]. This model predicts predominantly prompt (sub-picosecond) production of X-rays from clusters, which has not been observed in experiments (see chapter 7).

The nanoplasma model. The results of the models described above emphasise the importance of the electrons in the laser–cluster interaction. The high density inside a cluster (essentially the density of a solid) means that collisional ionisation can occur, which will enhance the rate of ionisation compared with that of individual atoms in low-density targets. The results of the Coulomb-explosion models and particle codes suggest that electrons can be retained in the cluster by the space charge of the ions. The electric field of the laser beam in the cluster will then be modified due to the presence of the free electrons.

These ideas are incorporated in the numerical model of the interaction which was developed by Ditmire *et al.* at Lawrence Livermore National Laboratory [9]. This model treats the exploding cluster as a small, spherical plasma – a 'nanoplasma'. For the cluster to be treated as a plasma, it must be larger than a Debye length,

$$\lambda_{\mathrm{D}} = \left(\frac{k_{\mathrm{B}} T_{\mathrm{e}}}{4\pi n_{\mathrm{e}} e^2} \right)^{\frac{1}{2}},\tag{6.1}$$

where n_{e} is the electron density, T_{e} the electron temperature, e the electronic charge and k_{B} the Boltzmann constant. For a solid-density Xe plasma ionised to 5^+ with an electron temperature of 1 keV, $\lambda_{\mathrm{D}} \approx 0.8$ nm, which corresponds to a 40-atom Xe cluster. So the plasma description is valid for clusters significantly larger than this. The model also assumes that the electric field is uniform across the cluster and that all the atoms in the cluster simultaneously experience the same electric field strength of the laser. This approximation is valid because the clusters considered are much smaller than the laser wavelength, λ, and the skin depth of the plasma,

$$\delta = \frac{c}{\omega_{\mathrm{p}}},\tag{6.2}$$

where c is the speed of light and ω_{p} is the plasma frequency in the nanoplasma, given by

$$\omega_p = \left(\frac{4\pi e^2 n_e}{m_e} \right)^{\frac{1}{2}}, \quad \omega_p = 5.64 \times 10^4 \sqrt{n_e \, [\text{cm}^{-3}]}, \qquad (6.3)$$

where m_e is the mass of the electron. For example, a very large cluster of 4×10^6 Xe atoms has a radius of 39 nm, 1/20th of the laser wavelength at 780 nm, and the skin depth drops to $\lambda/10$ only when the electron density rises to $100 n_{\text{crit}}$, where n_{crit} is the critical electron density given by

$$n_{\text{crit}} = \frac{\pi c^2 m_e}{e^2 \lambda^2}, \quad n_{\text{crit}} \, [\text{cm}^{-3}] = \frac{1.1 \times 10^{21}}{(\lambda \, [\mu\text{m}])^2}. \qquad (6.4)$$

It is also assumed that the temperature distribution across the plasma is isotropic, since the plasma is small enough and collisional enough that there are no temperature gradients. The ion-density distribution is assumed to be uniform across the cluster and the expansion is assumed to be self-similar, so the density remains uniform across the cluster throughout the expansion. The electron-energy distribution is assumed to be Maxwellian throughout.

The initial ionisation in the cluster is assumed to take place through tunnel ionisation [37]. After ionisation, a significant proportion of the electrons is retained within the cluster by the space charge of the ions. This means that both thermal and laser-driven collisional ionisation (inverse *Bremsstrahlung*) [38] occur within the cluster. The model does not include any estimate of rates of inner-shell excitation. Ionisation is assumed to be strictly sequential and no recombination is allowed to occur.

The heating of the cluster occurs primarily through collisional heating of the electrons. The rate of collisional heating inside the cluster is obtained by considering the cluster to be a uniform dielectric sphere in a uniform electric field. The presence of the free electrons means that the field inside the cluster can be shielded or enhanced relative to the external laser field. The field inside a dielectric sphere in a uniform electric field, E, is related to the vacuum electric field E_0 by [39]

$$E = \frac{3 E_0}{|\epsilon + 2|}. \qquad (6.5)$$

The dielectric constant ϵ is calculated using the Drude model [40]:

$$\epsilon = 1 - \frac{n_e / n_{\text{crit}}}{1 + i\nu/\omega}. \qquad (6.6)$$

Here ω is the laser frequency and ν the frequency of electron–ion collisions calculated from the Silin formulae [41]. The field inside the cluster is shielded from the external field when $n_e/n_{crit} > 6$ and, importantly, reaches a peak relative to the external electric field when

$$n_e = 3n_{crit}, \tag{6.7}$$

the so-called Mie resonance. At this point, the laser field resonantly drives the electron cloud of the cluster. As a result the rate of collisional heating of the free electrons in the cluster is greatly enhanced.

The expansion of the cluster is driven by two mechanisms: the build-up of charge on the cluster (the Coulomb force) and the hydrodynamic pressure from hot electrons which expand outwards, dragging the ions with them. The hydrodynamic pressure due to the electrons is given by

$$P_e = n_e k_B T_e. \tag{6.8}$$

The Coulomb pressure from the charge Qe on a sphere of radius r is

$$P_{Coul} = \frac{Q^2 e^2}{8\pi r^4}. \tag{6.9}$$

Since the electrons are largely retained within the cluster and the electron temperature is so high, the nanoplasma model predicts that the expansion is primarily due to the hydrodynamic pressure of the hot electrons rather than the Coulomb pressure of the ions. The hydrodynamic pressure scales as r^{-3} (through the electron density n_e) while the Coulomb pressure scales as r^{-4}. This means that the Coulomb pressure is expected to be relatively more important for smaller clusters. The rate of expansion is calculated by equating the rate of change of the kinetic energy of the cluster to the rate at which work is done by the plasma in its expansion [42].

The build-up of charge on the cluster is due to electrons freely streaming from the surface of the sphere. For this to occur, the electrons must be less than one mean free path [43] (in the direction of travel) from the surface and have an energy greater than the energy required for escape from the cluster. The escape energy is calculated from the potential energy on the surface of a sphere with charge Q and it is assumed that the charge is distributed isotropically over the sphere.

The electron and ion temperatures are calculated by accounting for the change of temperature due to expansion of the cluster, electron–ion equilibration [43] and heating of the cluster electrons. In a hydrodynamic expansion, random thermal energy is converted to directed kinetic energy, resulting in a change in temperature.

Computationally, the laser pulse is typically divided into 50 000 time steps. At each time step, the rates of tunnel, laser-driven and thermal ionisation are calculated, giving the number of ions in each charge state.

The total charge on the cluster after free-streaming of electrons is computed, together with the radius and radial velocity of the cluster. The new number densities of ions and electrons are then calculated and the ion and electron energies can then be found. All electrons with energies greater than the escape energy and within one mean free path of the surface are allowed to leave the cluster (and the energy distribution in the cluster re-thermalises). The energy distribution of the electrons which have left the cluster is thus the sum of the hot tails of the Maxwellian electron-energy distributions.

The model does not account for the lowering of the potential barriers due to the presence of neighbouring charged ions, as in the ionisation-ignition model. However, the numerical results indicate that the final energies of explosion are relatively insensitive to the initial rate of ionisation. The electron density quickly rises to the point at which the field inside the cluster is shielded from the laser and the ionisation is dominated by thermal collisional ionisation.

The calculated temporal history of an exploding cluster of 5000 Xe atoms irradiated by a 200-fs, 780-nm laser pulse with a peak intensity of 10^{16} W cm^{-2} is illustrated in figure 6.1. The peak of the laser pulse is at $t = 0$ fs. Early on in the laser pulse, at around $t = -280$ fs when the intensity is $\simeq 4 \times 10^{13}$ W cm^{-2}, a small number of free electrons are created through tunnel ionisation. The electron density rises to reach $3n_{crit}$ at $t = -270$ fs, see figure 6.1(c). At this point the field in the cluster is enhanced, see figure 6.1(a), and more electrons are liberated through tunnel, laser-driven and thermal ionisation. The electron density is now higher than $3n_{crit}$ and the field inside the cluster is shielded from the external laser field. The rates of tunnel and laser-driven ionisation fall off, but electrons are still created through thermal collisions.

From $t = -50$ fs onwards, some electrons are able to leave the cluster, for the mean electron temperature is in the region of 100–1000 eV and the escape energy is $\simeq 200$–2000 eV. The combined effect of the free-streaming of electrons out of the cluster and the hydrodynamic expansion of the cluster is that the electron density starts to fall, after peaking at over $50n_{crit}$. The field in the cluster again starts to rise as the electron density drops, so the rates of tunnel and laser-driven ionisation increase while the rate of thermal ionisation falls. Near the peak of the laser pulse, at $t = -12$ fs, the electron density in the cluster drops to $3n_{crit}$. The resonantly increased rate of heating causes the electron temperature in the cluster to soar to 25 keV, see figure 6.1(b). The field in the cluster is also strongly enhanced and the peak intensity in the cluster reaches 2×10^{16} W cm^{-2}, twice the intensity outside. The rate of free-streaming of electrons increases sharply, since a significant number of electrons has energies above the escape energy of 4 keV.

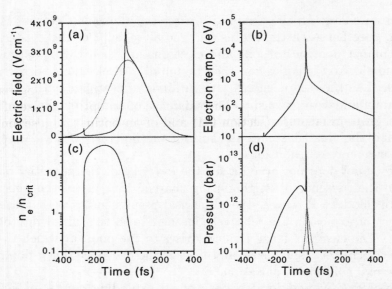

Fig. 6.1. The calculated temporal history of the explosion of a 5000-atom Xe cluster irradiated by a 200-fs, 780-nm laser pulse with a peak intensity of 10^{16} W cm^{-2}. (a) The envelope of the laser pulse (dotted line) and the field inside the cluster (solid line). (b) The electron temperature inside the cluster. (c) The electron density/critical electron density. (d) The hydrodynamic (solid line) and Coulomb (dotted line) pressures.

The total charge on the cluster increases to $5.5 \times 10^4 e$, resulting in the Coulomb pressure increasing to 10 Mbar, see figure 6.1(d). However, this is small relative to the hydrodynamic pressure due to the hot electrons of 200 Mbar. This pressure causes a sharp increase in the velocity of expansion of the cluster as the cluster starts to explode. Once the density of the nanoplasma has dropped to $\simeq 10^{17}$ cm^{-3} – the typical background density in an extended cluster-gas medium – the final velocity of expansion is 3.3×10^7 cm s^{-1}, which corresponds to a maximum ion energy of 79 keV. The final electron energy is much lower, only 30 eV (c.f. free-streaming energies of 0.2–2 keV), a consequence of their much smaller mass.

Discussion. The nanoplasma model makes several important predictions and experimental verifications of these are described in the following sections. These include energetic electrons (section 6.3), extremely high energies of ions (section 6.4) and charge states significantly higher than those obtained from low-density targets irradiated at the same intensity (section 6.4). The model predicts a characteristic two-lobed electron-energy spectrum, with a high-energy (a few kilo-electron-volts) peak of electrons ejected at the $3n_{crit}$ point as well as lower-energy electrons that leave the cluster earlier during the laser pulse. Finally, the model indicates that the explosion is driven by the resonant heating which occurs when the electron density drops to three times the critical electron density. The maximum ion temperatures occur when this point falls close to the peak of the laser pulse, leading to an optimum size of cluster for a given pulse duration and peak intensity. The existence of an optimum size of cluster for a given laser pulse (section 6.4) and of an optimum pulse length for a given size of cluster [18] are key pieces of evidence in support of the model.

Experimental: the Imperial College set-up

In this section, the set-up used at Imperial College London for the experimental investigation of explosions of isolated clusters is described. By way of an overview, the Imperial College laser is a high-power ($\simeq$0.3 TW), femtosecond titanium-doped-sapphire (Ti : sapphire) system, operating at a repetition rate of 10 Hz. The laser pulses interact with a low-density, skimmed cluster beam produced in the adiabatic expansion of high-pressure gas into vacuum. Ion and electron energies and charge states of ions are measured in a high-vacuum time-of-flight (TOF) spectrometer with a microchannel-plate (MCP) detector. The spectrometer can be configured with a variety of extraction or retardation fields to enable the different types of measurements.

The laser system. The femtosecond CPA Ti : sapphire laser has been described in detail in [44]. The laser produces femtosecond pulses of energy up to $\simeq$40 mJ centred at a wavelength of $\simeq$780 nm. The Kerr-lens mode-locked oscillator produces near-transform-limited pulses of duration $\simeq$90 fs that are amplified in the rest of the system by a factor of $\simeq$10^8. The oscillator pulses are stretched in a diffraction-grating stretcher to a duration of $\simeq$250 ps, thus reducing the pulse power by a factor of $\simeq$3000. This permits safe amplification first to the millijoule level in a Ti : sapphire regenerative amplifier operating at 10 Hz and then to $\simeq$80 mJ in a multi-pass Ti : sapphire power amplifier. The amplified pulses are then recompressed in a grating compressor to yield $\simeq$40-mJ, 150-fs pulses ($\simeq$0.3 TW) at a repetition rate of 10 Hz. In conjunction with standard (single-shot) autocorrelation pulse-duration measurements and equivalent-plane focal-spot characterisations, the focused laser intensity is inferred from ion-appearance data using over-the-barrier ionisation thresholds [45]. The ion-appearance data are recorded in the TOF spectrometer with a low-pressure filling of the desired atomic gas.

The particle spectrometer. The TOF spectrometer is illustrated in figure 6.2. To examine the interaction of intense laser pulses with isolated clusters (i.e. the response of single clusters), it is necessary for the laser to interact with a relatively low density of clusters to minimise the interaction between clusters and to reduce the effects of space charge which can distort the energy spectra of electrons and ions leaving the interaction volume. To achieve this, a high-density cluster jet produced from a solenoid-pulsed valve at the top of the chamber is skimmed to produce a beam of clusters of much lower density. The beam of clusters intersects with the laser focus at the centre of the high-vacuum section of the chamber. Typically, the density of clusters is $\simeq 10^{10}$ cm^{-3}, corresponding to $\simeq 10^4$ clusters in the focal volume. The linearly polarised laser is focused at $\simeq f/10$ with a plano-convex lens, which gives an approximately Gaussian focal spot of e^{-2} radius $\simeq 20$ μm. Angular distributions in the plane perpendicular to the laser beam are measured by varying the polarisation vector with a half-wave plate.

The energies of ions produced in the interaction are determined from their times of flight from the laser focus to the MCP detector without the use of an extraction field. The electron energies are found by measuring the decrease in MCP signal as a function of a retarding voltage applied to a grid placed between the focus and the MCP. In the same way as for the ions, no field is used to extract the electrons from the interaction region.

Measurements of sizes of clusters. At Imperial College, Rayleigh scattering measurements are used to establish the presence of clusters in the gas jet and to determine their average size from previously measured scaling rules [46, 47]. The following size scalings are determined for Xe and Kr as a function of the backing pressure, p_0:

$$N_c(\text{Xe } 300 \text{ K}) \approx 100(p_0 \, [\text{bar}])^2,$$
$$N_c(\text{Kr } 300 \text{ K}) \approx 30(p_0 \, [\text{bar}])^2. \tag{6.10}$$

The radii of clusters, R_c, proportional to $(N_c)^{1/3}$, are calculated assuming that the clusters have the same density as the bulk solid. This gives

$$R_c(\text{Xe}) \approx 2.39(N_c)^{1/3} \text{ Å},$$
$$R_c(\text{Kr}) \approx 2.26(N_c)^{1/3} \text{ Å}. \tag{6.11}$$

It has to be stressed that this optical scattering technique yields only information on the average size of the clusters; it does not provide any information on the size distribution of the clusters. Also, the error in N_c is quite large, probably a factor of two.

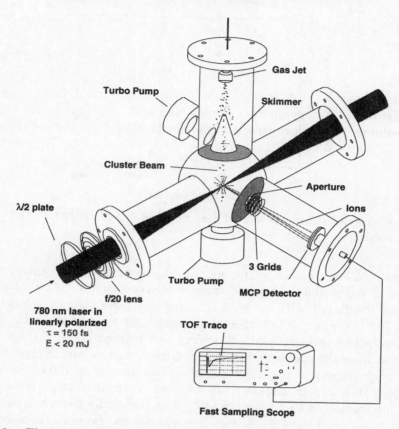

Fig. 6.2. The time-of-flight spectrometer used to measure electron and ion energies and charge states of ions from exploding clusters in the Imperial College experiments.

6.3 Electrons emitted from explosions of clusters

The production of extremely energetic electrons (>1 keV) in the laser–cluster interaction had been inferred ever since the first X-ray spectra from cluster plasmas were recorded (see discussion in section 6.2). The first measurement of the electron-energy spectrum from the explosion of isolated clusters exposed to intense laser pulses was made by Shao *et al.* at Imperial College [10]. This relatively early work, which is still the only one concerned with direct measurements of the energies of electrons emitted from clusters, played a key role in confirming the validity of the nanoplasma model for the explosion of large clusters heated by femtosecond laser pulses and also shed new light on the interaction dynamics.

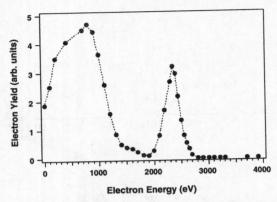

Fig. 6.3.   Electron energy spectrum from the explosion of 2100-atom Xe clusters at an intensity of 2×10^{16} W cm^{-2}.

Imperial College results. The measured electron energy spectrum is shown in figure 6.3. This spectrum was recorded for ≃2100-atom Xe clusters irradiated with ≃150-fs Ti:sapphire pulses at an intensity of ≃2×10^{16} W cm^{-2}. The spectrum is striking for two reasons. First, the production of electrons with energy out to nearly 3 keV is radically different from the spectra one would see from interactions of laser beams with single Xe atoms, for which energies of no more than 100 eV would be expected [48]. In fact, the electrons are even hotter than those for a laser–solid interaction at this intensity [49]. This highlights the remarkably efficient coupling of the laser energy into the cluster. Second, the spectrum has two peaks. The first, referred to as the 'warm' electrons, comprises electrons with energies in the range 0.1–1 keV. A sharper 'hot'-electron peak is also seen at an energy of ≃2.5 keV. Warm and hot electrons are produced under different conditions at different times during the expansion of a cluster. This conclusion was reached by examining the angular dependence of the electron emission with respect to the polarisation of the laser beam (figure 6.4). The measured angular distribution of warm electrons is shown in part (a) and that of the hot electrons in part (b) of this figure.

The angular distributions of the two groups of electrons are markedly different. The hot-electron emission is isotropic, while the warm-electron emission is peaked along the polarisation of the laser beam, with a full width at half-maximum of about 60°. These are completely different from the angular distributions associated with single atoms. The electrons from high-order above-threshold ionisation have a much narrower angular distribution (a width of 15–20°, peaked along the polarisation of the laser beam, was reported in [50]). In high-field tunnelling ionisation, the narrow angular distribution stems from the much higher tunnelling rate in the

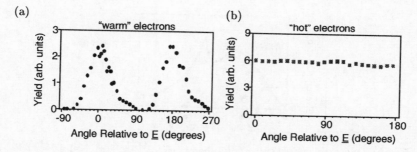

Fig. 6.4. (a) The angular distribution of 'warm' electrons, i.e. those with energies ranging from 0.3 to 1 keV. Emission along the polarisation of the laser beam's electric field is defined as 0° and 180°. (b) The angular distribution of 'hot' electrons, i.e. those with energies between 2 and 3 keV.

direction of the laser field. The electrons emitted from the clusters cannot be interpreted as simply resulting from the tunnel ionisation of individual atoms.

Discussion. By running simulations with their numerical nanoplasma code, the Imperial College researchers were able to reproduce the prominent features of the characteristic electron spectrum. Figure 6.5 shows the electron-energy spectrum calculated using values of parameters close to the experimental ones. A broad warm peak and a sharper hot peak near 2.5 keV are seen, in agreement with the experimental data. The agreement is relatively good, but there are discrepancies. For example, the peaks in the calculated spectrum are broader. This difference is most probably due to the assumption that the electron distribution within the cluster is Maxwellian (which greatly simplifies the calculations). In reality, the distribution is not Maxwellian; the hottest electrons in the outer tail of the distribution leave the cluster first and the fast disassembly of the cluster prevents complete thermalisation by electron–electron collisions.

The nanoplasma model provides considerable insight into the origin of the two electron features. It shows that the warm-electron peak is the result of collisional heating of electrons near the surface of the cluster on the rising edge of the laser pulse. The hot electrons result from rapid heating of the remaining electrons in the bulk of the cluster later during the pulse when the electron density drops to such a level as to bring the heating into resonance. This explanation is consistent with the observed angular distribution data. The warm electrons have undergone a limited number of collisions, broadening the angular distribution from that of purely tunnel-ionised electrons. The hot electrons, on the other hand, result from extensive collisional heating of the electrons in the bulk of the cluster. Consequently, their velocity distribution has been randomised

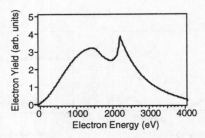

Fig. 6.5. An electron spectrum calculated by using the Imperial College nanoplasma code.

completely, accounting for the isotropic distribution observed.

From this investigation it is clear that electron spectroscopy provides extremely useful information for understanding the laser–cluster dynamics.

6.4 Ions emitted from explosions of clusters

Overview. Around the same time as the cluster-electron measurements were made, Snyder *et al.* at Penn State University reported some interesting results concerning emission of ions from Xe clusters irradiated by femtosecond laser pulses at an intensity of $\simeq 10^{15}$ W cm^{-2} [14]. Xe ions were being produced with charge states as high as 18^+, much higher than would be produced from single Xe atoms interacting with such laser pulses. Snyder's group proposed that these ions were being produced by ionisation ignition (see section 6.2) followed by Coulomb explosion.

At a laser intensity about an order of magnitude higher, the Imperial College team of Ditmire *et al.* then made some startling experimental measurements, which could not be explained in terms of the prevailing Coulomb-explosion model. They recorded *average* energies of $\simeq 45$ keV for ions being produced in the interaction of intense femtosecond-laser pulses with clusters of several thousand Xe atoms [11, 12]. The details of this investigation are provided below.

The production of energetic ions. Raw ion-TOF spectra obtained from the interaction of $\simeq 2500$-atom Xe clusters with a laser pulse of intensity $\simeq 2 \times 10^{16}$ W cm^{-2} (accumulated over several thousand laser shots) are shown in figure 6.6(a). By varying the half-wave plate in the beam path before the focusing lens, it was found that the ion-energy distribution is almost isotropic with respect to the direction of polarisation of the laser beam – see the curves in figure 6.6(a) – a clear signature of a spherically symmetric explosion of a cluster. The corresponding ion-energy spectrum is displayed in part (b) of this figure.

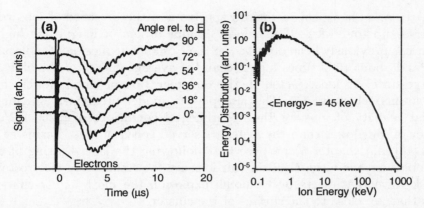

Fig. 6.6. (a) Ion-TOF spectra from the explosion of 2500-atom Xe clusters irradiated at 2×10^{16} W cm^{-2} for a range of angles of emission (the distance of flight was 0.38 m). (b) The corresponding energy spectrum. The mean energy is 45 keV and some of the ions have energies greater than 1 MeV.

The abrupt unresolved feature in the TOF spectrum is due to the electrons (warm and hot). The mean energy of ions in the spectrum in figure 6.6(b) is 45 ± 5 keV, showing that the average energy deposited by the laser beam per ion is substantial (this is the explanation for the very-high-energy absorption exhibited by dense cluster gases – see chapter 7 for details). Bear in mind that the ions produced from the high-intensity irradiation of single Xe atoms would have essentially zero kinetic energy at this laser intensity. The hydrodynamic model offers a simple explanation for the high average energies of ions measured for the explosions of Xe clusters. In the ambipolar expansion of the cluster both electrons and ions ultimately reach a velocity given roughly by the speed of sound in the cluster plasma,

$$c_\text{s} = \sqrt{ZkT_\text{e}/m_\text{i}}, \tag{6.12}$$

where Z is the charge state of the ion, kT_e is the thermal energy of the electron and m_i is the ion's mass. Most of the resulting kinetic energy is, however, contained in the ions due to their much greater mass. The average energy of ions will therefore be of the order of

$$\tfrac{1}{2}m_\text{i}c_\text{s}^2 \approx ZkT_\text{e}. \tag{6.13}$$

This implies, for example, that the average energy of Xe ions will be $\simeq 50$ keV if we assume that the electron temperature is given by the hot-electron feature in figure 6.3, i.e. $kT_\text{e} \approx 2.5$ keV, and the average charge state is $Z \approx 20^+$ (see the next section). This is in good agreement with the mean energy of $\simeq 45$ keV observed.

A remarkable aspect of the ion spectrum is the presence of ions with energies up to 1 MeV. This energy is four orders of magnitude higher than has previously been observed in the Coulomb explosion of molecules [51] and about 1000 times higher than the average energy of the highest-charge-state Ar ions ejected in the disintegration of small (fewer than ten atoms) clusters [7]. The nanoplasma model predicts a maximum radial velocity of ions for the experimental conditions of $\simeq$250 keV, a factor of four lower than the fastest observed speed of ions. Lezius *et al.* (see [16] and chapter 5) propose that, following the laser heating of the nanoplasma, the mega-electron-volt ions are the result of an electrostatic shock wave that ejects by Coulomb explosion the most highly charged ions that are close to the surface of the cluster.

Dobosz *et al.* [19] recently proposed a quite different mechanism for the production of mega-electron-volt ions. They suggest that standing waves in the electric field can be set up between the over-critical cluster nanoplasmas and that, in these standing waves, unclustered ions are ponderomotively accelerated to mega-electron-volt energies. Further tests to establish the validity of these mechanisms are necessary.

Charge-state distributions. The production of highly stripped ions is another characteristic feature of the laser–cluster interaction, which has already been observed for laser–cluster interactions at lower intensity [5, 8, 14]. The Imperial College workers measured the charge states of ions from single-cluster explosions by placing three closely spaced metal grids in the flight tube. Charging the middle grid to a potential Φ while keeping the front and back grids at earth introduced a barrier to ions with energies less than $Ze\Phi$, without significantly altering the times of flight of higher-energy ions. By varying the voltage they were able to measure the charge-state distribution of the ions as a function of kinetic energy [11, 12].

The charge-state data in figure 6.7 were obtained for $\simeq$2500-atom Xe clusters interacting with Ti : sapphire laser pulses at a peak intensity of $\simeq 2 \times 10^{16}$ W cm^{-2}. For high-energy Xe ions (>100 keV), the peak charge state is $Z = 18^+$ to 25^+, with some ions, remarkably, having charge states as high as 40^+. These are much higher than the $\simeq 12^+$ expected from field ionisation of single atoms at these intensities [52]. In fact, ionisation to Xe^{40+} would require an intensity of nearly 10^{20} W cm^{-2} [45] if the ionisation were due to over-the-barrier field ionisation. The nanoplasma model provides an explanation for these high charge states. High-temperature electrons in the cluster, which are created through laser-driven heating, strip the ions to higher charge states by collisional ionisation. The charge state of an ion depends only weakly on its kinetic energy, contrary to what would be expected from a pure Coulomb explosion and contrary to the

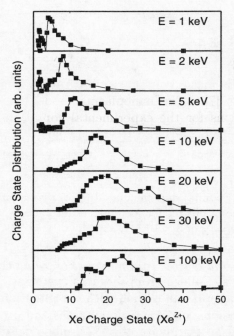

Fig. 6.7.   Measured charge-state distributions from the explosion of 2500-atom Xe clusters irradiated at 2×10^{16} W cm^{-2} for kinetic energies of 1, 2, 5, 10, 20, 30 and 100 keV per ion.

results reported in [7] for the Coulomb explosion of very small clusters containing Ar.

Using a magnetic-deflection TOF spectrometer, Lezius *et al.* at Saclay [16] have subsequently gone a step further and recorded the energy distributions of charge states in the range up to 30$^+$ from 2×10^6-atom Xe clusters interacting with 130-fs Ti:sapphire laser pulses at an intensity of $\simeq 5 \times 10^{17}$ W cm^{-2}. They could thus investigate the dependence of the measured ion energies upon the charge state of ions. Their results suggest that the Xe clusters exhibit an expansion that has both a Coulomb explosion and hydrodynamic character, with the highest energies of ions arising from the Coulomb contribution.

In terms of applications, a table-top source of highly stripped, heavy ions is of great interest to the particle-accelerator community [53]. To produce the large ion fluxes needed, it is envisaged that the laser must interact with a high-density cluster target. A still-unanswered question is that of whether, under these conditions, for which space charge and recombination are important effects, the highly charged ions from explosions of clusters are able to leave the volume of interaction [54].

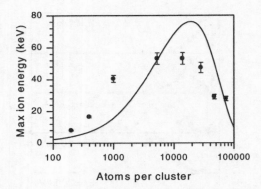

Fig. 6.8.   Measured (points) and calculated (line) maximum energies of ions as a function of the size of cluster for Xe clusters interacting with 230-fs, 780-nm pulses at a peak intensity of 3×10^{15} W cm^{-2}.

Scalings with laser and cluster parameters. Following these initial investigations, the Imperial College group began a systematic survey of parameters, to test the nanoplasma model, with the intention of optimising the energy of the explosions, e.g. for fusion studies and production of X-rays (see chapters 5 and 7). Springate *et al.* [22] made detailed measurements of how the energies of ions vary with the cluster and laser parameters, for rare-gas clusters interacting with 200-fs pulses from the Ti:sapphire laser system.

Scaling with the size of cluster. Figure 6.8 shows how the maximum energy of ions varies with the number of atoms in a Xe cluster in the range 200–74 000 Xe atoms. A laser beam of intensity 3×10^{15} W cm^{-2} was used. The threshold size for production of energetic (>1 keV) ions appears to be $\simeq$200–400 atoms. The maximum energy of ions rises from 8 keV at 200 atoms per cluster to a peak of $\simeq$53 keV for clusters of $\simeq$8000 atoms before falling to $\simeq$28 keV as the size is increased beyond 50 000 atoms per cluster.

Numerical calculations from the nanoplasma model are also shown in figure 6.8 (solid line). The model predicts the existence of an optimum size of $\simeq$20 000 atoms per cluster. The predicted variation in energy of ions with size is in good agreement with that measured experimentally. The nanoplasma model suggests that the dynamics of the explosion is governed by the point in the laser pulse at which the cluster experiences the resonant heating. The highest ion temperatures are obtained when the cluster passes through the $3n_{crit}$ point (see equation (6.7)) near the peak of the laser pulse. Small clusters expand more quickly and reach this point before the peak; larger clusters pass through $3n_{crit}$ well after the peak. Increasing the intensity increases the initial rates of ionisation

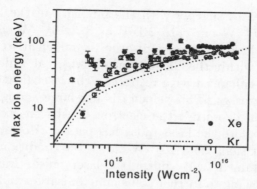

Fig. 6.9. Maximum energies of ions measured as a function of the intensity of the laser beam for 5300-atom Xe clusters (solid points) and 6200-atom Kr clusters (open points), together with the simulated scalings for Xe clusters (solid line) and Kr clusters (dotted line).

and expansion so that the $3n_{crit}$ point occurs earlier in the laser pulse for a given size of cluster, thus shifting the optimum size towards larger clusters. The confirmation of an optimum size for a given set of laser-pulse parameters, together with the existence of an optimum pulse width for a given size of cluster [9, 18], is strong evidence for a resonance in the cluster heating.

Scaling with the intensity of the laser. The scaling of the energies of ions with the intensity of the laser beam has also been investigated both for Xe and for Kr clusters by the Imperial College group [22] (figure 6.9). The Xe clusters contained $\simeq 5300$ atoms and the Kr clusters $\simeq 6200$ atoms. Similar behaviour is observed for these two ion species, with $\simeq 20\%$ lower energies for Kr. In Xe, a sharp onset of hot-ion production at an intensity of 6×10^{14} W cm^{-2} is seen. The ion energies rise steeply up to $\simeq 10^{15}$ W cm^{-2}, for which the maximum energy of ions is $\simeq 50$ keV. At intensities, I, higher than these, the maximum energy increases only as $\simeq I^{0.2}$ while the integrated yield of ions increases as $\simeq I^{1.4}$, which is consistent with the increase in focal volume (an $I^{3/2}$ scaling would be expected for a Gaussian focal spot). At 10^{16} W cm^{-2}, the maximum energy is 90 keV. The energies of Kr ions follow a similar trend to that for Xe, with a sharp increase up to a mean energy of 9 keV at 10^{15} W cm^{-2} followed by a slow increase up to energies of 75 keV at 10^{16} W cm^{-2}.

For comparison, the calculated intensity dependence of the mean energy attained by ions in the explosion is also shown in figure 6.9 (solid line for Xe, dotted line for Kr). The nanoplasma model predicts that Xe and Kr will exhibit very similar scalings with intensity for the laser and cluster parameters used in the experiment. For both ion species, the energy increases sharply with the intensity of the laser beam from 40 eV at 6×10^{13} W cm^{-2} to 2 keV at 3×10^{14} W cm^{-2}. In this intensity region, the amount of ionisation is sufficient to ensure that the electron density within the cluster reaches $3n_{\mathrm{crit}}$, at which point the field in the cluster is enhanced with respect to the laser field and the velocity of expansion increases. This velocity is not large enough relative to the rate of ionisation to permit the electron density in the cluster to drop to $3n_{\mathrm{crit}}$ again before the laser pulse has passed. However, when the intensity is greater than $\simeq 6 \times 10^{14}$ W cm^{-2}, the condition of enhancement of heating is attained twice during the laser pulse. First, on the rising edge and then, owing to further collisional ionisation and expansion, near the peak of the pulse, at which point the field in the cluster is strongly enhanced and the radial acceleration increases dramatically. In this intensity regime the energies of ions increase only slowly with increasing intensity (as $\simeq I^{0.5}$) because, as the intensity increases, the amount of initial ionisation increases. This means that the cluster expands faster and the electron density drops through $3n_{\mathrm{crit}}$ earlier in the laser pulse. The cluster does not then experience a substantially higher intensity at the point of resonant heating.

The numerical results are in good agreement with the measurements, matching the general trends and predicting the intensity regime within which the energies of ions increase most rapidly. Although the scaling of energy with intensity is fairly weak from 10^{15}–10^{16} W cm^{-2}, the model predicts that the energies will start to rise more steeply at intensities above 3×10^{16} W cm^{-2}.

Scaling with the laser wavelength. The wavelength scaling of cluster dynamics has been examined by several groups in the context of yields of X-rays from dense cluster targets. Kondo and co-workers [55, 56] found that the yield from clusters was much lower at 800 nm than it was at 248 nm. On the other hand, Dobosz *et al.* [57] found that there was no strong scaling of the yield on going from the first to the second harmonic of their laser. Springate *et al.* at Imperial College have examined the scaling with laser wavelength of single-cluster explosion dynamics by carrying out a comparative study of the ion-energy spectra at the fundamental and the second harmonic [22]. For these experiments the Ti : sapphire laser was frequency doubled to 390 nm in a KDP crystal. Figure 6.10 shows the variation of the maximum energies of ions with the size of Xe cluster for 780-nm (infra-red) and 390-nm (blue) pulses at nominally the same intensity. ($\simeq 4 \times 10^{15}$ W cm^{-2}). Numerical calculations comparing the energies of ions emitted from clusters irradiated with the two wavelengths are also presented and it is apparent that the agreement between experiment and theory is poor. Except for the largest clusters, the ions produced with

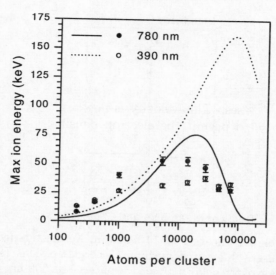

Fig. 6.10. Measured (points) and calculated (lines) maximum energies of ions emitted from the explosion of Xe clusters with 780-nm (solid points, solid lines) and 390-nm (open point, dotted lines) laser pulses.

the infra-red pulse are of higher energies (by up to a factor of two) than occur with the blue pulse. In contrast, the nanoplasma model predicts that, at sizes of cluster below $\simeq 10\,000$ atoms, the mean energy in the blue is $\simeq 5$ keV higher with 390-nm irradiation than it is with irradiation at 780 nm. The energies obtained with the second harmonic are predicted to be much higher than those with the fundamental for large clusters. The optimum size of cluster is calculated to be $\simeq 20\,000$ atoms at 780 nm and $80\,000$ atoms at 390 nm. Both experiment and the model agree that the optimum size is larger in the blue than it is in the infra-red.

At the present time, it is not clear whether the poor agreement between theory and experiment reflects a deficiency in the nanoplasma code or whether it is an experimental artefact, e.g. arising from the pulse shapes or pre-pulse levels for the laser fundamental and second harmonic being different. Further investigations, perhaps also examining the electron energies, are required in order to resolve this uncertainty.

6.5 Increasing the energy of explosions

Owing to the efforts made by a number of groups, understanding of the laser–cluster interaction has improved to the point at which strategies for boosting the energy of explosions can be considered. The ultimate aim is to optimise the conditions in a bulk cluster plasma for a particular application, such as fusion or production of X-rays.

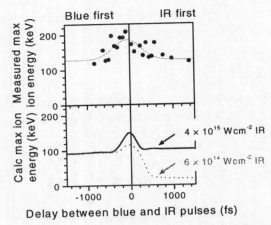

Fig. 6.11. Double-kick heating of 14 000-atom Xe clusters using 390-nm and 780-nm laser pulses. Measured maximum energies of ions (solid points) are shown as a function of the delay between the pulses for a 390-nm pulse of intensity 3×10^{15} W cm^{-2} and a 780-nm pulse of intensity 6×10^{14} W cm^{-2}. The dotted line is a Gaussian fit to the data, centred at -45 fs. The solid and dashed lines are calculations for a 390-nm, 185-fs, 3×10^{15}-W cm^{-2} pulse and a 780-nm, 260-fs pulse at 4×10^{15} W cm^{-2} (solid line) and 6×10^{14} W cm^{-2} (dashed line).

Two-colour heating. Because the $3n_{\mathrm{crit}}$ point is wavelength dependent (scaling as λ^{-2}, see equation (6.4)), the nanoplasma model predicts that a cluster irradiated with two precisely timed laser pulses of different wavelengths will experience enhanced heating twice during its expansion. To test this 'double-kick' hypothesis, Springate and co-workers at Imperial College have conducted an investigation into the heating of Xe clusters using two laser pulses [21], one at the fundamental (780 nm) and the other at the second harmonic (390 nm) generated in a KDP crystal. The delay between the pulses could be controlled to within $\simeq 30$ fs.

Figure 6.11 shows the mean energy of ions emitted from the explosion of $\simeq 14\,000$-atom Xe clusters as a function of the delay between the two pulses (a negative delay corresponds to the blue pulse preceding the infra-red pulse). The 780-nm and 390-nm pulses were of intensities 6×10^{14} and 2.5×10^{15} W cm^{-2}, respectively, chosen to give approximately the same mean energies of ions separately. As expected from the enhanced-heating picture, the optimum delay condition is found to be the blue pulse preceding the infra-red by $\simeq 40$ fs. The mean energy of ions is then about twice that obtained with either pulse alone. A similar enhancement was observed for the maximum energy of ions. The intensity scaling of the energies also indicates that the energy obtained for the optimum delay is greater than if the energy in both laser pulses were combined into a single pulse at either frequency.

To rule out the possibility that the enhancement of the heating is simply a consequence of the increase in the peak electric field of the laser beam when the pulses are overlapped, Springate *et al.* compared the ion energies when the two pulses were polarised parallel and perpendicular to each other. No significant difference was observed, supporting the two-kick picture and not consistent with ionisation ignition (see section 6.2) as a model of the laser–cluster interaction, which should be strongly dependent on the peak electric field, since it depends on the enhancement of tunnel ionisation in the cluster due to the proximity of charged ions in the clusters.

Mixed-species clusters. Mixed-species clusters (i.e. clusters containing two or more elements, such as clusters of molecular gases) offer two important advantages over single-species clusters. Firstly, they potentially permit elements that are difficult to make cluster in the pure form to participate in the energetic interaction described in the preceding sections. For example, formation of large clusters in gas jets of hydrogen (and its isotopes for fusion experiments) requires very high backing pressures and cryogenic cooling, because the condensation parameter for hydrogen is very low (see table 7.1 in chapter 7). Hydrogen iodide (HI) gas, on the other hand, forms large clusters very readily at low pressures without any cooling.

The second advantage of mixed-species clusters is that a high average charge state, Z, for ions in the cluster is required for efficient collisional heating of the plasma electrons. According to the nanoplasma model, the maximum energy of ions emitted from the explosion of single-species clusters of the same number of atoms scales as Z^2. Therefore, in a cluster comprising light and heavy elements, the more highly stripped heavier ion species should allow a higher electron temperature, T_e, to be attained to drive the hydrodynamic expansion of both ion species. In the expansion of two-species clusters, another process can come to play, which can further boost the energies of the ions. For expansions of two-species planar plasmas, it is known that an 'impurity' species (having a charge–density product much smaller than that of the other ion species) can attain a much higher energy in the expansion than can the other species, with a small fraction of these ions achieving energies of up to $\simeq 1000 T_e$ [58].

To investigate some of these intriguing possibilities, Tisch *et al.* at Imperial College performed an experimental study of the laser heating of large HI clusters [20]. This molecule, which clusters very readily, was chosen because a deuterated version (DI) would be relevant to fusion research on clusters (see chapter 7) and because the presence of iodine (which has nearly the same atomic number as Xe) in the cluster should allow efficient laser heating of the cluster nanoplasma. Further, the large difference in

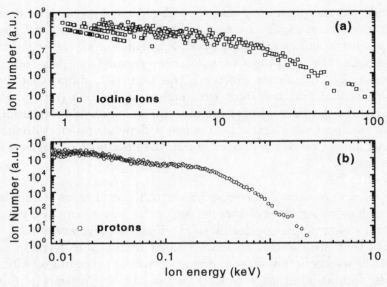

Fig. 6.12.   Measured ion spectra from explosion of 60 000-molecule HI clusters irradiated by 260-fs, 780-nm pulses at a peak intensity of 2×10^{16} W cm^{-2}: (a) the iodine ion spectrum and (b) the proton spectrum.

charge state between H and I allows H to be treated as an impurity ion, despite its having an equal number density to I in the nanoplasma.

HI clusters were formed in the adiabatic expansion of high-pressure HI gas using the same pulsed valve and skimmer arrangement as that employed for formation of rare-gas clusters. Using a backing pressure of $\simeq 8$ bars, HI clusters of an estimated mean size of 60 000 molecules were produced. These were irradiated by 260-fs Ti:sapphire pulses at an intensity of 2×10^{16} W cm^{-2}. By carrying out pulse-height analysis of the MCP signal (detailed in [20]) it was possible to distinguish between impacts of protons and iodine ions on the MCP and thus record separate energy spectra, as shown in figures 6.12(a) and (b), respectively. The maximum kinetic energy of iodine ions is $\simeq 92$ keV, while the mean of the distribution is $\simeq 8.6$ keV. The mean and maximum kinetic energies of protons are approximately 270 eV and 2.5 keV, respectively. A monotonic increase was observed for energies both of I^{n+} and of protons with increasing size of HI cluster up to 60 000 molecules per cluster, which was the largest size that could be generated owing to technical limitations.

The observation of energetic protons and iodine ions proves that molecular clusters can indeed be used to produce energetic ions from a range of elements much broader than those that cluster readily in the pure form. Detailed quantitative analysis of the ion spectra is, however, difficult owing to the lack of a mixed-species laser–cluster model at the present time.

Though Tisch *et al.* did not measure ion spectra from pure H or I clusters in this experiment, comparison with explosions of single-species clusters can be made from their nanoplasma simulations. The maximum energy of iodine ions emitted from the explosion of HI clusters is lower than that predicted for a pure I cluster with the same number of iodine atoms. This can be attributed, at least in part, to the lower effective Z in the HI nanoplasma owing to half the ions being only singly charged. The maximum energy of protons appears to be approximately the same as for a single-species (H cluster) explosion. However, since it is much easier to make large HI clusters rather than pure H clusters (cryogenic cooling not being required), DI clusters may be a more practical choice for cluster fusion.

Discussion. The Imperial College researchers could not determine unambiguously whether any protons had been accelerated to the very high energies predicted for an impurity ion in the two-species expansion, owing to the presence of a masking feature in the TOF spectrum. It is worth noting that a small fraction of very fast deuterons might have a significant effect on the fusion yield, given the very strong scaling of the fusion cross section with the deuteron energy. Clearly, further experimental work is necessary in order to establish whether this impurity acceleration does, in fact, occur in the explosion of a two-species cluster. This is an area of laser–cluster research in which experiment is leading theory, since the existing nanoplasma codes are inadequate for describing clusters of more than element. The development of a mixed-species-cluster model – perhaps drawing on work done on two-species planar laser plasmas [59] – will help to assess the potential of such clusters for boosting the explosion energy of low Z-ions.

6.6 High-harmonic generation in clusters

Background

In this section we deviate a little from the theme of interactions of laser beams with isolated clusters to discuss a different aspect of the laser–cluster interaction that, although it is still in its infancy, appears to be a fertile area for future research. This is the generation of coherent vacuum-ultra-violet radiation and soft X-rays from a medium of atomic clusters through the process of high-order-harmonic generation (HHG). Though HHG in clusters involves the interaction of an intense laser pulse with a relatively high-density cluster medium, we shall see that mechanism for generation of harmonics in clusters and the optical properties of clusters at high laser intensity are connected to the physics governing the explosion of isolated clusters, i.e. collective behaviour of electrons is implicated.

HHG in atomic media (and to a lesser extent molecular media) has been studied in great detail over the last decade [60] and a good understanding both of the single-atom emission and of the coherent growth of the harmonic field in a macroscopic medium (known as 'phase-matching') has been attained. A semi-classical model of the single-atom emission [61, 62], based on acceleration of the electron in the laser field and recollision with the ion, has proved successful for describing certain aspects of this phenomenon and reproducing the important features of numerical simulations based on solution of the time-dependent Schrödinger equation (see [27] and references therein). A simple scaling rule has emerged from the semi-classical picture for the highest harmonic photon energy possible [61]:

$$E_{\text{max}} = I_{\text{p}} + 3U_{\text{p}}(I_{\text{sat}}), \qquad (6.14)$$

where I_{p} is the ionisation potential of the atom (or ion) and $3U_{\text{p}}(I_{\text{sat}})$ is the maximum energy gained by the electron accelerating in the laser field before returning to the ion. Here, U_{p} is the ponderomotive potential given by

$$U_{\text{p}} = \frac{e^2 E^2}{4m_{\text{e}}\omega^2}, \quad U_{\text{p}}\,[\text{eV}] = 9.33 \times 10^{-14} I\,[\text{W cm}^{-2}]\,(\lambda\,[\mu\text{m}])^2, \qquad (6.15)$$

where E is the electric field of the laser, I is the intensity of the laser and ω and λ are its frequency and wavelength, respectively. The intensity used in equation (6.14) is the intensity at which ionisation saturates, I_{sat}, since, above this intensity, HHG is terminated for that species.

The HHG spectrum has recently been extended into the 2–4-nm 'water window' [63, 64] – of great interest for biological imaging – corresponding to harmonic orders in excess of 300. High-harmonic radiation is directional, has a high degree of temporal and spatial coherence [65] and has a pulse duration as least as short as that of the driving laser [66], making it a unique table-top, ultra-fast, short-wavelength source. However, the efficiency of the process is extremely low (especially below 10 nm, where the efficiency of energy conversion is typically $<10^{-13}$ per harmonic order), owing primarily to the dispersion of free electrons produced through ionisation of the medium that breaks the phase-matching condition required for coherent growth of the harmonic field [60].

HHG in cluster jets

Overview. Cluster targets have recently come under scrutiny as non-linear media for HHG. Certainly, on the face of it, atomic clusters possess a number of properties making them very interesting non-linear media for HHG. The collective behaviour of electrons seen in explosions of clusters suggests that collective phenomena may also affect the effective dipole involved in the production of harmonics. In addition, cluster electrons see a binding potential that is modified from the single-atom case by the close proximity of neighbouring ions. Also, bound and free charges in the cluster can result in shielding of the laser field inside the cluster. Finally, the linear optical properties of a cluster medium at high laser intensity that determine the phase-matching conditions for HHG are quite different from those of a partially ionised atomic medium, having dispersion characteristics (see section 6.6) that might permit the phase-matching of very-high-order harmonics to be improved dramatically.

The first report of HHG from clusters came from Donnelly and co-workers at Lawrence Livermore, who reported the generation of harmonics up to the 31st order of a 140-fs, 825-nm laser in the 10^{14}–10^{15} W cm^{-2} range from 1000-atom Ar clusters produced in a high-pressure gas jet [67]. For comparison, harmonics were also generated in a gas cell containing Ar in which no clusters were present. They found that the harmonics generated in clusters displayed a slightly increased appearance intensity, a phenomenon attributed to shielding of the laser field by plasma electrons inside the cluster. Their data also suggested that the cluster harmonics exhibited a stronger non-linear dependence on the intensity than did those generated in the static-gas cell, a possible manifestation of the modified binding potential in the cluster.

HHG in xenon clusters. Following the Livermore work, an investigation of HHG in Xe clusters was undertaken at Imperial College [68], employing their femtosecond Ti : sapphire laser. Tisch and co-workers attempted to make a direct comparison of the harmonic yields from clusters and atoms, which could not be done accurately in Donnelly's experiment because of the difficulty of comparing the density distributions in their clustered gas jet and gas cell. To compare the intensity of harmonics from clusters and atoms at the same average density, they used a 'weak' nanosecond ultra-violet laser pulse ($\simeq 10^{10}$ W cm^{-2}) to disassemble the clusters into their constituent atoms ahead of the main laser pulse. Rayleigh scattering of green Nd : YAG laser light was used to monitor formation and disassembly of clusters. Concentrating on the harmonic-wavelength range 80–90 nm and relatively low laser intensities (10^{13}–10^{14} W cm^{-2}) the Imperial College physicists found that HHG in Xe clusters was up to five time more efficient than that in atomic Xe at the same average particle density. The yield data for the fifth harmonic of the 390-nm second harmonic of the Ti : sapphire laser is shown in figures 6.13(a) and (b) for backing pressures of 1 and 4 bars Xe, respectively.

The 1-bar data reveal no difference between the harmonic yield with (solid points) versus without (open points) the disassembling pre-pulse.

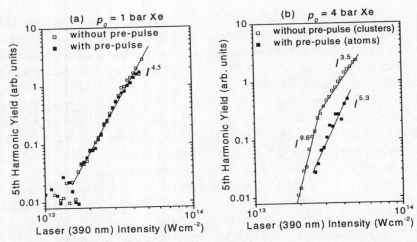

Fig. 6.13. A comparison of fifth-harmonic yields (at wavelength 78.0 nm) without (open squares) and with a pre-pulse (solid squares) as a function of laser intensity for backing pressures of (a) 1 bar Xe, before significant clustering, and (b) 4 bars Xe, when the medium comprises $\simeq 2500$-atom clusters in the absence of the pre-pulse.

This was consistent with previous cluster-jet characterisation, which showed that this backing pressure was below the onset of formation of large (>100 atoms) clusters for Xe. However, at 4 bars Xe clusters of average size $\simeq 2500$ atoms are formed and there is a marked increase (by up to a factor of five) in the harmonic yield without the pre-pulse, when the laser is interacting with Xe clusters. What is also apparent from figure 6.13(b) is that, as Donnelly observed, the harmonic yield from the clusters initially has a much stronger dependence on the laser intensity (scaling as $\simeq I^{10}$) than the $\simeq I^5$ scaling of the yield from the monatomic medium that is consistent with lowest-order perturbation theory.

When the Imperial College team measured the dependence of the harmonic yield on the gas-jet backing pressure they discovered an anomalous scaling, in comparison with the case for atomic media. This stimulated them to formulate a picture for the mechanism of enhancement of HHG in clusters based on a collective oscillation of electrons, somewhat akin to the Mie resonance responsible for the enhanced laser heating of cluster nanoplasmas. The pressure-scaling data for the ninth harmonic of the 780-nm Ti:sapphire laser are displayed in figure 6.14.

The top part of this figure shows that, up to a pressure of 1 bar, the harmonic yield scales as the square of the backing pressure, p_0. This is the expected n_0^2 scaling of coherent HHG (where n_0 is the atomic density in the region of interaction), bearing in mind that, before the onset of clustering, $n_0 \propto p_0$. However, at the onset of cluster formation at $\simeq 1$ bar

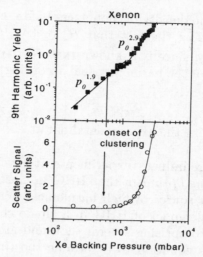

Fig. 6.14. Top: the ninth harmonic yield (at wavelength 86.7 nm) as a function of the backing pressure of Xe (laser intensity $\simeq 5 \times 10^{13}$ W cm^{-2}). Bottom: Rayleigh-scattering data from the gas jet show that the change in slope of the harmonic pressure dependence to $\simeq p_0^3$ coincides with the onset of clustering at $p_0 \approx 1$ bar.

– clearly shown by the sharp increase in the Rayleigh-scattering signal from the target (bottom part of figure 6.14) – the scaling changes to $\simeq p_0^3$. As a control, they also measured the pressure scaling in He, which does not cluster in this pressure range. The expected p_0^2 scaling for atoms was found.

The simple picture proposed by the Imperial College researchers that reproduces the approximately cubic backing-pressure scaling of the harmonic yields is based on a collective oscillation of the cluster charge. In this model, the electron cloud formed from the optically active electrons in the cluster oscillates about the positively charged ion matrix. It is assumed that this gives rise to a global cluster dipole at the harmonic frequency, $d_c(q\omega)$, that increases, in analogy with an atomic dipole, proportionally to the cluster's charge multiplied by its radius, i.e.

$$d_c(q\omega) \propto N_c R_c. \tag{6.16}$$

Assuming that the phase-matching does not change significantly as the size of cluster is increased (more on this assumption later), the scaling for the qth harmonic yield from the clusters is thus given by

$$N_q \propto n_c^2 |d_c(q\omega)|^2 \propto n_c^2 N_c^2 R_c^2, \tag{6.17}$$

where n_c is the density of clusters. Further assuming that all the atoms emerging from the nozzle condense into clusters, it follows that $n_c =$

n_0/N_c, where n_0 is the density of monomers before clustering. Recalling that $n_0 \propto p_0$ and using the fact that $R_c \propto (N_c)^{1/3}$, this means that $N_q \propto p_0^2 N_c^{2/3}$. Finally, since it is known that $N_c \propto p_0^2$ [46], the backing-pressure dependence of the harmonic yield according to this global-dipole picture is

$$N_q \propto p_0^{10/3}, \tag{6.18}$$

in good agreement with the experimental data.

Discussion. These preliminary experiments at Livermore and Imperial College allow for some optimism that HHG in cluster media could lead us to a more-efficient source of high harmonics. What is lacking now is a detailed theoretical model of HHG in a single cluster. For the large clusters of interest, comprising several thousand atoms, numerical solution of Schrödinger's equation to determine the time-dependent dipole moment, as is routinely performed for simplified atomic systems, is out of the question.

However, such calculations have been performed for much smaller clusters by Hu and Xu [35, 36]. They have solved the two-dimensional Schrödinger equation with multiple potential wells in two dimensions, in an attempt to simulate small clusters (<50 atoms) in which the active electron of an atom in the cluster is subject to the combined field of its parent ion and all the other charges in the cluster. Their calculations show that, for a single active electron in a 49-well potential, the yields for harmonics in the 50–100 nm range are increased by up to two orders of magnitude over those for an isolated atom. However, models based on the dynamics of a single active electron in the cluster environment are unable to simulate collective oscillations of electrons that may lead to the formation of a global cluster dipole. Obviously further theoretical work to shed more light on this intriguing topic is required.

Phase-matching in cluster media

Background. Implicit in the comparison between clusters and monomer media described above was the assumption that the phase-matching is the same in both cases, allowing the differences observed to be attributed to the response of a single particle (atom or cluster). While this assumption might be valid for a relatively low laser intensity, recent theoretical work by a few groups suggests that, for laser intensities high enough to produce a cluster nanoplasma, the linear optical properties of a cluster medium that govern the phase-matching conditions may differ significantly from those of an atomic medium.

For maximum efficiency of HHG, the wavenumbers of the harmonic field and its source, the non-linear polarisation at the harmonic frequency, must be equal. In practice, several factors result in a finite wavenumber mismatch, Δk, that prevents the achievement of this 'phase-matched' condition. Considering only the material dispersion of the generating medium (the most problematic source of Δk for HHG), the phase-matching condition $\Delta k = 0$ requires that the refractive indices at the frequencies of the laser (ω_1) and the qth harmonic ($q\omega_1$) be identical, i.e.

$$n(\omega_1) = n(q\omega_1). \tag{6.19}$$

However, to achieve high harmonic orders, a laser intensity close to that for the saturation of ionisation of the non-linear medium must be used. Under these conditions, Δk is large, being dominated by the dispersion of free electrons in the region of interaction. The free-electron dispersion is given by

$$n(\omega) \approx \sqrt{1 - N_e/n_{crit}(\omega)}, \tag{6.20}$$

where N_e is the *free*-electron density (not to be confused with n_e, the electron density inside the cluster nanoplasma) and n_{crit} is given by equation (6.4). Several schemes have been suggested in order to improve the phase-matching conditions for HHG. They include quasi-phase-matching [69] and phase-matching in gas-filled hollow fibres [70], which has been demonstrated down to $\simeq 20$ nm in weakly ionised media. However, for wavelengths below 10 nm, including the important water-window range, the free-electron dephasing is still viewed as the fundamental limitation to the conversion efficiency for HHG.

Buffering HHG with clusters. The first hint that clusters might provide an elegant new solution to this long-standing problem can be found in the work of Kim *et al.* [71]. In order to investigate theoretically the propagation of intense, femtosecond laser pulses in a cluster jet, they derived an expression for the refractive index of a partially ionised cluster medium, based on a nanoplasma picture of the individual clusters. Later, Tajima *et al.* [72] published a paper containing a theoretical examination of the optical properties of a gas of ionised clusters. They formulated a dispersion relation for the laser–cluster interaction, assuming *static* cluster parameters, and showed that the unique dispersion properties of the ionised cluster medium could be harnessed to phase-match HHG.

Meanwhile, Tisch had been looking at the frequency dependence of the refractive index formulated by Kim *et al.* in order to investigate the phase-matching conditions in a cluster medium interacting with very intense laser pulses. Using the nanoplasma code he was able to incorporate into his phase-matching model realistic *time-dependent* cluster parameters to simulate the explosion of the clusters that would inevitably be occurring at the high laser intensities being considered [73]. Independently corroborating and extending the work of Tajima *et al.*, Tisch proposed a scheme for improving the phase-matching of HHG in the presence of ionisation using the dispersion of a gas of atomic clusters in a manner very reminiscent of the 'buffer-gas' method well known in low-order nonlinear optics [74].

A gas of cluster nanoplasmas (a 'cluster plasma') is the first medium that has been identified that can be used as a buffer gas at *high* laser intensities at which significant ionisation is occurring. In fact, it is only at high intensity that the cluster gas has the desired properties. The cluster plasma has a Mie resonance around which the sign of its linear susceptibility changes and this characteristic variation in susceptibility can be used to compensate for the normal dispersion ($\mathrm{d}n/\mathrm{d}\omega > 0$) of the free electrons. The time-dependence arises from the extremely rapid temporal variation not only of the cluster parameters that change the position of the Mie resonance (equation (6.7)) on the femtosecond timescale but also of the free-electron density from tunnel ionisation of the monomer medium (numerical modelling shows that the clusters do not add significantly to the free-electron density, since the cluster electrons are largely confined to the nanoplasma in the time window under consideration [75].)

In the scheme envisaged by Tisch [73], HHG in a monomer rare-gas plasma is buffered through the addition of a cluster gas. Normally, HHG from plasma ions is prohibited owing to the strong free-electron dispersion, described above. However, by buffering the interaction with clusters, the full potential of ions for HHG might be realised. This potential can be seen from equation (6.14); ions possess a higher I_{p} than that of the neutral species and consequently a larger I_{sat}. Thus the ionic dipole can be driven much harder before ionisation, resulting in the generation of higher-order harmonics than can occur with the neutral atom.

Tisch chose to examine a high-density medium comprising Ar clusters and monomeric He. A single femtosecond-laser pulse at a peak intensity of $\simeq 10^{16}$ W cm^{-2} is used both to ionise the medium to create the required plasma and to generate high-order harmonics in a novel travelling-wave phase-matching scheme in which the buffered region propagates with the laser and harmonic pulses.

Figure 6.15 shows the results of dispersion calculations for a 30-fs, 800-nm laser pulse at a peak intensity of 10^{16} W cm^{-2} interacting with the mixture of He atoms and Ar clusters. The density of He is 3.85×10^{18} cm^{-3} (chosen to optimise the phase-matching), while the density of Ar clusters is 2.25×10^{16} cm^{-3} with a realistic log-normal size distribution with a mean size of 840 atoms, which is also the width of the distribution [76]. The time-dependent coherence length for the harmonic $q = 265$ (at a wavelength of 3 nm which is in the middle of the water window) is plotted for the Ar–He mixture (solid curve) and for He alone (dashed curve). The

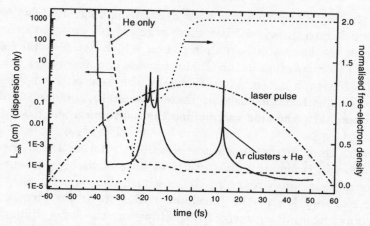

Fig. 6.15. Calculated coherence lengths (considering dispersion only) for generation of the 265th harmonic driven by a 30-fs, 800-nm laser pulse at a peak intensity of 10^{16} W cm^{-2}, for He alone (solid line – refer to the left-hand axis) and for the same density of He mixed with Ar clusters (dashed line – refer to the left-hand axis). The dotted line (refer to the right-hand axis) is the calculated free-electron density normalised with respect to the initial density of He. The dot–dashed line is the normalised laser-intensity profile.

coherence length, defined as

$$L_{\mathrm{coh}}(t) = \pi/\Delta k(t), \qquad (6.21)$$

is that length over which the harmonic field grows coherently and so sets an upper limit on the useful interaction length. The harmonic-conversion efficiency, in the absence of saturation, scales as the square of L_{coh}. (Note that, since only the free-electron dispersion is included, the coherence lengths are initially infinite when the electron density is zero.) The dotted curve is the normalised free-electron density arising from tunnel ionisation of He, while the dot–dash curve is the laser pulse.

For the pure He medium, $L_{\mathrm{coh}}(t)$ drops rapidly as the free-electron density increases, flattening out at a value of $\simeq 0.8$ μm near the peak of the laser pulse. This is much shorter than the millimetre-scale interaction lengths available, highlighting the inefficiency of HHG in an ionised medium. For the mixture of Ar clusters and He, $L_{\mathrm{coh}}(t)$ exhibits two main peaks where the phase-matching is dramatically improved, owing to the buffering action of the clusters. Only the first peak is of relevance for HHG, since the second phase-matching peak occurs after the saturation of ionisation of He$^+$ (i.e. fully stripped He). Averaging over the three very narrow (<1 fs) sub-peaks, the width of the peak is $\simeq 5$ fs and the peak's coherence length is $\simeq 1$ mm, comparable to the maximum high-density interaction length possible in a standard gas jet. During this peak, generation of the 265th harmonic from the He^+ ions is phase-matched over the interaction length, provided that there is negligible group-velocity walk-off between the laser and harmonic pulses, which Tisch found to be the case [73]. The dispersion in the cluster plasma can also buffer the phase-mismatch arising from the Gouy phase of the focused laser (which has the same sign as the free-electron phase-mismatch), provided that it has an approximately constant value along the interaction length.

Discussion. Only further work will establish whether these theoretical predictions will translate into a practical phase-matching scheme in the laboratory. One uncertainty in the cluster-buffering scheme is that of whether reabsorption of the harmonic radiation in the cluster plasma will become the limiting factor [19]. However, the large enhancements predicted (an increase by about a factor of 10^2 in coherence length when other sources of phase-mismatch are included [73]) that would unleash a wide range of scientific and technological applications of high-harmonic radiation should provide a strong motivation for investigating these ideas experimentally.

6.7 Future directions

Although there have been significant developments in the last five years in the understanding of the interaction of intense, femtosecond-laser pulses with single clusters, notably, the success of the nanoplasma model, much basic research has still to be done, both theoretically and experimentally.

On the theoretical front, the nanoplasma model of the interaction needs further development. Although it has proved remarkably useful, in a few respects its predictions are not in good agreement with experimental data. The scaling of the interaction with the wavelength of the laser beam, the mechanism of production of the most energetic ions and the details of the expansion phase of the nanoplasma are all areas in which there are discrepancies between the model and experiment, but the sources of the discrepancies are not clear. These areas require further theoretical investigation.

As highlighted in preceding sections, for certain aspects of the laser–cluster interaction, e.g. mixed-species clusters and collective behaviour of electrons contributing to the single-cluster harmonic dipole, there are currently no theoretical models. These are inherently difficult subjects that pose considerable challenges to theoreticians working in the field.

There are many avenues to be explored experimentally, many of which have been touched upon in the preceding sections. Priority must be given to more accurate cluster-size measurements and, importantly, measurement of the size distribution of clusters in the interaction volume. Once the cluster-size distribution has been obtained, it will be easy to incorporate it into the modelling. Other topics for future work involving single-cluster explosions include high-energy (>10 keV) electron spectroscopy, aimed at measuring the very-high-energy electrons that are predicted to undergo free-streaming at the Mie resonance; tests for impurity acceleration of low-Z ions in mixed-species laser–cluster interactions; control of cluster-explosion dynamics through adaptive tailoring of the laser pulses to optimise cluster-plasma conditions for applications; and detailed wavelength scaling of the energetics of explosion of clusters.

The optical properties of cluster media seems to be a very fertile area for future experimental study. Is the global cluster-dipole picture valid? Laser ellipticity studies might be used to test this, since the extreme sensitivity of the HHG yields to the ellipticity seen in single atoms [77] should be reduced in the giant-dipole picture, wherein recollision with the parent ion is not necessary. Are the coherence properties of harmonics emitted from clusters the same as those of harmonics emitted from atoms? Can HHG be phase-matched at very high laser intensity in a mixture of monomer and cluster gases? As well as their predicted buffering action in HHG, the unusual linear optical properties of cluster media might also be harnessed in the future for a range of other applications, including acceleration of electrons [72] and the guiding of intense laser pulses [71] that have already been suggested. Many questions remain to be answered. One certainty is that this will be an area of growth in laser–cluster research.

Acknowledgements

The Imperial College results presented in this chapter are due to the concerted efforts of a dedicated team of researchers and support staff in the Blackett Laboratory Laser Consortium (BLLC). The authors would like to thank T. Ditmire, A. Gregory, E. Gumbrell, N. Hay, H. Hutchinson, K. Mendham, J. Marangos, M. Mason, P. Ruthven, Y. Shao and R. Smith for their invaluable contributions. The BLLC programme is supported by the UK Engineering and Physical Sciences research council (EPSRC) and J. Tisch is supported by an EPSRC Advanced Fellowship.

References

[1] E. W. Becker, K. Bier and W. Henkes, *Z. Phys.* **146** 333 (1956).

[2] C. Brechignac, P. Cahuzac, J. Leygnier and A. Sarfati, *Phys. Rev. Lett.* **70** 2036 (1993).

[3] J. Bregnacac and J. P. Connerade, *J. Phys. B* **27** 3795 (1994).

[4] A. McPherson, T. S. Luk, B. D. Thompson, K. Boyer and C. K. Rhodes, *App. Phys. B* **57** 337 (1993).

[5] A. McPherson, B. D. Thompson, A. B. Borisov, K. Boyer and C. K. Rhodes, *Nature* **370** 631 (1994).

[6] A. McPherson, T. S. Luk, B. D. Thompson, A. B. Borisov, O. B. Shiryaev, X. Chen, K. Boyer and C. K. Rhodes, *Phys. Rev. Lett.* **72** 1810 (1994).

[7] J. Purnell, E. M. Snyder, S. Wei and A. W. Castleman Jr, *Chem. Phys. Lett.* **229** 333 (1994).

[8] T. Ditmire, T. Donnelly, R. W. Falcone and M. D. Perry, *Phys. Rev. Lett.* **75** 3122 (1995).

[9] T. Ditmire, T. Donnelly, A. M. Rubenchik, R. W. Falcone and M. D. Perry, *Phys. Rev. A* **53** 3379 (1996).

[10] Y. L. Shao, T. Ditmire, J. W. G. Tisch, E. Springate, J. P. Marangos and M. H. R. Hutchinson, *Phys. Rev. Lett.* **77** 3343 (1996).

[11] T. Ditmire, J. W. G. Tisch, E. Springate, M. B. Mason, N. Hay, R. A. Smith, J. Marangos and M. H. R. Hutchinson, *Nature* **386** 54 (1997).

[12] T. Ditmire, J. W. G. Tisch, E. Springate, M. B. Mason, N. Hay, J. P. Marangos and M. H. R. Hutchinson, *Phys. Rev. Lett.* **78** 2732 (1997).

[13] E. M. Snyder, S. Wei, J. Purnell, S. A. Buzza and A. W. Castleman Jr, *Chem. Phys. Lett.* **248** 1 (1996).

[14] E. M. Snyder, S. A. Buzza and A. W. Castleman Jr, *Phys. Rev. Lett.* **77** 3347 (1996).

[15] M. Lezius, S. Dobosz, D. Normand and M. Schmidt, *J. Phys. B* **30** L251 (1997).

[16] M. Lezius, S. Dobosz, D. Normand and M. Schmidt, *Phys. Rev. Lett.* **80** 261 (1998).

[17] T. Ditmire, E. Springate, J. W. G. Tisch, Y. L. Shao, M. B. Mason, N. Hay, J. P. Marangos and M. H. R. Hutchinson, *Phys. Rev. A* **57** 369 (1998).

[18] J. Zweiback, T. Ditmire and M. D. Perry, *Phys. Rev. A* **59** R3166 (1999).

[19] S. Dobosz, M. Schmidt, M. Perdrix, P. Meynadier, O. Gobert, D. Normand, K. Ellert, A. Ya. Faenov, A. I. Magunov, T. A. Pikuz, I. Yu. Skobelev and N. E. Andreev, *J. Exp. Theor. Phys.* **88** 1122 (1999).

[20] J. W. G. Tisch, N. Hay, E. Springate, E. T. Gumbrell, M. H. R. Hutchinson and J.P. Marangos, *Phys. Rev. A* **60** 3076 (1999).

[21] E. Springate, N. Hay, J. W. G. Tisch, M. B. Mason, T. Ditmire, J. P. Marangos and M. H. R. Hutchinson, *Phys. Rev. A* **61** 4101 (2000).

[22] E. Springate, N. Hay, J. W. G. Tisch, M. B. Mason, T. Ditmire, M. H. R. Hutchinson and J. P. Marangos, *Phys. Rev. A* **61** 3201 (2000).

[23] M. D. Perry and G. Morou, *Science* **264** 917 (1994).

[24] D. Strickland and G. Morou, *Opt. Commun.* **56** 219 (1985).

[25] P. Agostini and G. Petite, *Contemp. Phys.* **29** 57 (1988).

[26] K. Burnett, V. C. Reed and P. L. Knight, *J. Phys. B* **26** 561 (1993).

[27] M. Protopapas, C. H. Keitel and P. L. Knight, *Rep. Prog. Phys.* **60** 389 (1997).

[28] B. Sheehy and L. F. DiMaurio, *Ann. Rev. Phys. Chem.* **47** 463 (1996).

[29] K. Codling and L. J. Frasinski, *Contemp. Phys.* **35** 243 (1994).

[30] P. Gibbon and S. Forster, *Plas. Phys. Controlled Fusion* **38** 769 (1996).

[31] I. Last, I. Scheck and J. Jortner, *J. Chem. Phys.* **107** 6685 (1997).

[32] E. F. Rexer, R. L. DeLeon and J. F. Garvey, *J. Chem. Phys.* **107** 4760 (1997).

[33] T. Ditmire, *Phys. Rev. A* **57** R4094 (1998).

[34] C. Rose-Petruck, K. J. Schafer, K. R. Wilson and C. P. J. Barty, *Phys. Rev. A* **55** 1182 (1997).

[35] S. X. Hu and Z. Z. Xu, *Appl. Phys. Lett.* **71** 2605 (1997).

[36] S. X. Hu and Z. Z. Xu, *Phys. Rev. A* **56** 3916 (1997).

[37] M. V. Ammosov, N. B. Delone and V. P. Krainov, *Sov. Phys. JETP* **64** 1191 (1986).

[38] W. Lotz, *Z. Phys.* **216** 241 (1968).

[39] J. D. Jackson, *Classical Electrodynamics*, John Wiley & Sons, New York, 1975.

[40] W. L. Kruer, *Physics of Laser Plasma Interactions*, Addison-Wesley, New York, 1988.

[41] V. P. Silin, *Sov. Phys. JETP* **20** 1510 (1965).

[42] A. F. Haught and D. H. Polk, *Phys. Fluids* **13** 2825 (1970).

[43] L. Spitzer, *Physics of Fully Ionised Gases*, Interscience, New York, 1962.

[44] D. J. Fraser and M. H. R. Hutchinson, *J. Mod. Opt.* **43** 1055 (1996).

[45] S. Augst, D. D. Meyerhofer, D. Strickland and S. L. Chin, *Phys. Rev. Lett.* **63** 2212 (1989).

[46] R. Karnbach, M. Joppien, J. Stapelfeldt and J. Wörmer, *Rev. Sci. Instrum.* **64** 2828 (1993).

[47] J. Wörmer, V. Guzielski, J. Stapelfeldt and T. Möller, *Chem. Phys. Lett.* **159** 321 (1989).

[48] T. E. Glover, T. D. Donnelly, E. A. Lipman, A. Sullivan and R. W. Falcone, *Phys. Rev. Lett.* **73** 78 (1994).

[49] G. Güthlein, M. E. Foord and D. Price, *Phys. Rev. Lett.* **77** 1055 (1996).

[50] U. Mohideen, M. H. Sher, H. W. K. Tom, G. D. Aumiller, O. R. Wood, R. R. Freeman, J. Bokor and P. H. Bucksbaum, *Phys. Rev. Lett.* **71** 509 (1993).

[51] C. Cornaggia, M. Schmidt and D. Normand, *J. Phys. B* **27** L123 (1994).

[52] S. Augst, D. D. Meyerhofer, D. Strickland and S. L. Chin, *J. Opt. Soc. Am. B* **8** 858 (1991).

[53] J. Collier, G. Hall, H. Haseroth, H. Kugler, A. Kuttenberger, K. Lamgbein, R. Scrivens, T. Sherwood, J. Tambini, A. Shumshurov and K. Masek, *Laser Particle Beams* **14** 283 (1996).

[54] P. Fournier, H. Haseroth, H. Kugler, N. Lisi, R. Scrivens, F. V. Rodrigue, P. DiLazzaro, F. Flora, S. Duesterer, R. Sauerbrey, H. Schillinger, W. Theobald, L. Veisz, J. W. G. Tisch and R. A. Smith, *Rev. Sci. Instrum.* **71** 1405 (2000).

[55] K. Kondo, A. B. Borisov, C. Jordan, A. McPherson, W. A. Schroeder, K. Boyer and C. K. Rhodes, *J. Phys. B* **30** 2707 (1997).

[56] W. A. Schroeder, F. G. Omenetto, A. B. Borisov, J. W. Longworth, A. McPherson, C. Jordan, K. Boyer, K. Kondo and C. K. Rhodes, *J. Phys. B* **31** 5031 (1998).

[57] S. Dobosz, M. Lezius, M. Schmidt, P. Meynardier, M. Perdrix and D. Normand, *Phys. Rev. A* **56** R2526 (1997).

[58] A. V. Gurevich, L. V. Pariiskaya and L. P. Pitaevskii, *Sov. Phys. JETP* **36** 274 (1973).

[59] P. E. Young, M. E. Foord, A. V. Maximov and W. Rozmur, *Phys. Rev. Lett.* **77** 1278 (1996).

[60] A. L'Huillier, L. Lompré, G. Mainfray and C. Manus, in *Atoms in Intense Laser Fields*, ed. M. Gavrila, Academic Press, Boston, 1992, pp. 139–202.

[61] J. L. Krause, K. J. Schafer and K. C. Kulander, *Phys. Rev. Lett.* **68** 3535 (1992).

[62] P. B. Corkum, *Phys. Rev. Lett.* **71** 1994 (1993).

[63] Z. Chang, A. Rundquist, H. Wang, M. Murnane and H. C. Kapteyn, *Phys. Rev. Lett.* **79** 2967 (1999).

[64] M. Schnürer, C. Spielmann, P. Wobrauschek, C. Streli, N. H. Burnett, C. Kan, K. Ferencz, R. Koppitsch, Z. Cheng, T. Brabec and F. Krausz, *Phys. Rev. Lett.* **80** 3236 (1998).

[65] T. Ditmire, J. W. G. Tisch, E. T. Gumbrell, R. A. Smith, D. D. Meyerhofer and M. H. R. Hutchinson, *Appl. Phys. B* **65** 313 (1997).

[66] J. W. G. Tisch, D. D. Meyerhofer, T. Ditmire, N. Hay, M. B. Mason and M. H. R. Hutchinson, *Phys. Rev. Lett.* **80** 1204 (1998).

[67] T. D. Donnelly, T. Ditmire, K. Neuman, M. D. Perry and R.W. Falcone, *Phys. Rev. Lett.* **76** 2472 (1996).

[68] J. W. G. Tisch, T. Ditmire, D. J. Fraser, N. Hay, M. B. Mason, E. Springate, J. P. Marangos and M. H. R. Hutchinson, *J. Phys. B* **30** L709 (1997).

[69] P. L. Shkolnikov, A. Lago and A. E. Kaplan, *Phys. Rev. A* **50** R4461 (1994).

[70] C. G. Durfee III, A. R. Rundquist, S. Backus, C. Herne, M. M. Murnane and H. C. Kapteyn, *Phys. Rev. Lett.* **83** 2187 (1999).

[71] A. V. Kim, D. Anderson, M. D. Chernobrovtseva and M. Lisak, in *Super-strong Fields in Plasmas* (AIP Conf. Proc. 426) pp. 79–84 (1998).

[72] T. Tajima, Y. Kishimoto and M. C. Downer, *Phys. Plasmas* **6** 3759 (1999).

[73] J. W. G. Tisch, *Phys. Rev. A* **62** 041802(R) (2000).

[74] J. F. Reintjes, *Nonlinear Optical Parametric Processes in Liquids and Gases* pp. 49-52, Academic Press, Orlando, 1984.

[75] E. L. Springate, *Atomic Clusters in Intense Laser Fields*, Ph. D. Thesis, Imperial College, London (1999).

[76] M. Lewerenz, B. Schilling and J. P. Toennies *Chem. Phys. Lett.* **206** 381 (1993).

[77] K. S. Budil, P. Salieres, A. L'Huillier, T. Ditmire and M. D. Perry, *Phys. Rev. A* **48** R3437 (1993).

7

Interactions of intense laser beams with extended cluster media

Roland A. Smith

The Blackett Laboratory
Imperial College, Prince Consort Road
London SW7 2BQ, UK

Todd Ditmire

Lawrence Livermore National Laboratory, PO Box 808,
L-477, Livermore, California 94550, USA

7.1 Introduction

As outlined in earlier chapters, there is currently considerable interest in the interaction of intense, short-duration laser pulses with isolated atomic clusters. Some of the more spectacular recent results reported in the literature have included the observation of mega-electron-volt ions [1–3], multi-kilo-electron-volt electrons [4, 5] and extremely high charge states of ions [1, 2, 6] in laser–cluster-interaction experiments. It is clear that, whilst aggregates of more than a few hundred atoms in strong laser fields share some aspects of both solid and monatomic gas-target behaviours, they also exhibit new and quite startling effects that are very specific to atomic clusters. In particular the interaction is extremely energetic compared with that of small molecules and aggregates of a few tens of atoms, for which field and multi-photon ionisation dominate and ions are typically accelerated to energies of only a few tens of electron-volts in weak Coulomb explosions.

An important question to ask is 'does the energetic behaviour seen in single-cluster experiments hold for extended cluster media, in which a laser pulse interacts with many billions of clusters at the same time and interactions between adjacent clusters become important?'. Are there interesting propagation effects, for example, and can we enhance coherent processes such as high-harmonic generation in cluster media? Finally, can we harness the large energies of ions available from single-cluster interactions for use in heating significant volumes of material to drive applications such as the production of X-rays and thermonuclear fusion. In this chapter we review a number of methods for generating atomic clusters and ways of characterising the medium produced, with a bias towards the pulsed gas jets used in many recent experiments. We also discuss pro-

216

cesses such as absorption and transport of energy in bulk cluster media and highlight recent work that uses the large energies available in cluster-interaction experiments to drive other processes such as thermonuclear fusion.

7.2 High-density cluster sources

Over the years atomic clusters have been of interest to a number of disciplines for several quite different and distinct reasons. In the 1970s and early 1980s clusters were first investigated by the physical chemistry community [7] because the very low (millikelvin) temperatures attainable during formation of clusters could be used to make feasible a range of sensitive spectroscopic techniques for the study of molecular dynamics [8]. More recently the dynamics of small metallic clusters was the focus of attention as a method of probing the transition from isolated atoms to the collective solid state. With the advent of ultra-short-pulse, high-intensity laser systems and the use of clustered gas media in interaction physics, clusters are once again a hot topic. However, despite several decades of work the formation of clusters has still not been described well analytically and there still remain distinct problems in the production, characterisation and use of cluster media in experiments.

A good review of the formation of clusters in free-jet expansions is provided by Hagena [9]. Briefly, cluster formation can occur in any supercooled gas medium when the mean kinetic energy of the gas atoms is less than the energy of formation of inter-atomic or inter-molecular bonds. Clusters are seen to form in noble-gas systems in which the van der Waals bond energies are of order 10^{-3} eV [10], metallic systems in which bond energies are of order 0.5 eV [11] and in intermediate regimes in which hydrogen bonding or formation of molecular bonds can take place. In all these cases a vapour stream needs to be formed and then rapidly cooled. This cooling is generally achieved by rapid expansion of a gas stream into vacuum. Suitable supersaturated vapours and gases can be produced from a range of sources including gas jets [12–15] (see figure 7.1), effusive ovens [16] and laser-ablated solids [17]. A light gas such as He can also be added to the cluster stream to act as a carrier (particularly when high-Z metallic clusters are required [18, 19]) and can also help to remove momentum in the three-body interactions required for 'adhesive' collisions to take place in the gas stream.

While we concentrate here on the formation of clusters, it is worth noting that pulsed valves of the type employed in many cluster-interaction experiments have also been used extensively to produce dense jets of monatomic gases for investigations into high-harmonic generation, wake-field acceleration of particles, etc. As a result there are papers describing

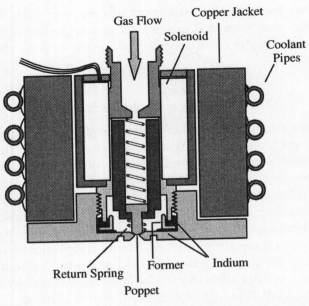

Fig. 7.1.   A cross section through a typical cooled solenoid valve showing details of the seal (a deformable ring of indium), movable poppet seal and solenoid assembly.

their performance in the literature [20, 21] that can give useful insights, methods for generating long interaction regions through the use of slit jets [22], etc.

7.3 The formation of clusters in free-jet expansions

Expansion of gas through a nozzle into vacuum results in substantial cooling in the frame of the moving gas, and atoms and molecules that interact weakly at room temperature can form clusters as a result. As outlined in chapter 5, the clustering is primarily determined by the temperature and pressure of the gas reservoir, the shape and size of the nozzle and the strength of the inter-atomic bonds formed [23, 24]. As is the case with many complex fluid-dynamics problems, there is currently no rigorous theory to predict formation of clusters in a free-jet expansion. However, the onset of clustering and size of clusters produced can be described by an empirical scaling parameter Γ^*, referred to as the Hagena parameter [24, 25]:

$$\Gamma^* = k \frac{(d/\tan\alpha)^{0.85}}{T_0^{2.29}} P_0, \tag{7.1}$$

Table 7.1. *Values of the Hagena parameter k for a range of gases from [24]†*
and calculated from [26]*

Gas	H_2	D_2	N_2	O_2	CO_2	CH_4
k	184*	181*	528*	1400*	3660*	2360†

Gas	He	Ne	Ar	Kr	Xe
k	3.85†	185†	1650†	2890†	5500†

where d is the nozzle diameter (millimetres), α the half-angle of expansion ($\alpha = 45°$ for sonic nozzles and $\alpha < 45°$ for supersonic nozzles), P_0 the backing pressure (millibars), T_0 the pre-expansion temperature (kelvins) and k a constant related to bond formation (see table 7.1 for a range of k values for common atomic and molecular gases).

Clustering typically begins for values of $\Gamma^* > 100–300$, with the number of atoms per cluster N_c scaling as $N_c \propto \Gamma^{*\,2.0–2.5}$ [23, 27]. We can thus use the strong variation of N_c with T_0 and P_0 to engineer a cluster medium of arbitrary mean density and N_c, for interaction experiments. However, this also requires detailed knowledge of the behaviours of the valve and cluster jet over a broad range of conditions. Fortunately this is somewhat simplified by the fact that the model of corresponding jets [9] allows the properties of a jet measured for one gas to be usefully extrapolated to other gases. However, it is also important to point out that there can be quite significant differences in the performance of a given pulsed gas jet depending on how it has been set up.

The cluster-size scaling derived from the Hagena parameter refers to the average number of atoms per cluster. However, nucleation and aggregation of clusters in a cooling gas is a statistical process and hence gives rise to a distribution of sizes. This distribution has not been studied very much and typically is assumed to follow a log-normal relationship [28] of the type

$$P_N(N) = \left(N\sigma\sqrt{2\pi}\right)^{-1} \exp\left[-(\ln N - \mu)^2/(2\sigma^2)\right], \qquad (7.2)$$

where μ and σ are the mean and standard deviation of the distribution of the natural logarithm of the size of clusters. It is difficult to establish a relationship between the observed log-normal size distribution and the dynamics of the formation of clusters because distributions of this type occur both during growth and during fragmentation [29].

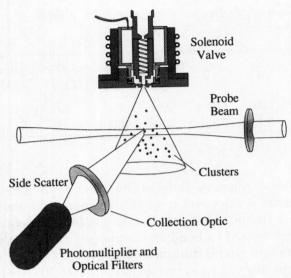

Fig. 7.2. A simple experimental arrangement for cluster-size characterisation using Rayleigh scattering. Replacing the PMT by a CCD imaging system at 90° to the probe beam allows the spatial profile of the cluster jet to be investigated.

7.4 Characterising cluster-size distributions

Changes of several orders of magnitude in the peak energy of ions and efficiency of absorption of laser light in a cluster gas occur as the size N_c is increased from a few tens to a few thousands of atoms [1–3, 5, 6] and so a good estimate of N_c is critical in interpreting interaction experiments. However, the small size and relative fragility of atomic clusters make accurate absolute measurements of size and size distributions rather difficult. Over the years several methods have been used to try to address this problem and there is a significant body of literature dealing with this field. With the advent of numerical models that make predictions for strong-field behaviour of clusters sensitive to size effects of less than a factor of two [30] and high-quality experimental data to compare against simulations, addressing this problem has once again become topical and important.

Optical scattering measurements

One of the simplest cluster-characterisation techniques is optical scattering. Clusters are substantially smaller than the wavelength of visible light and so scattering takes place in the classical Rayleigh regime. The total signal S_{RS} scattered from a cluster medium is proportional to the product of the scattering cross section σ and the number density n_c of the clusters,

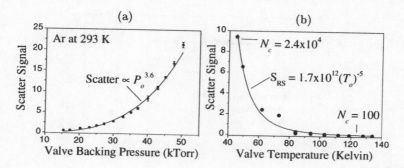

Fig. 7.3. (a) The peak 90°-scattered light signal as a function of pressure for Ar at 293 K, showing the onset of scattering and a near cubic dependence on the backing pressure, taken from [12]. (b) The size of D_2 clusters as a function of the valve temperature determined by Rayleigh-scattering measurements, taken from [12].

with the classical Rayleigh cross section [31] given by

$$\sigma = \frac{8\pi}{3} \frac{r^6}{\lambda^4} \left(\frac{n^2 - 1}{n^2 + 2} \right)^2, \tag{7.3}$$

with r the cluster's radius, λ the optical wavelength and n the refractive index of the scattering medium. Unlike Mie scattering (which becomes important for $r > 0.1\lambda$), Rayleigh scattering is independent of angle and the signal scattered from a moderately dense (10^{18} atoms cm^{-3}) clustered gas jet is relatively easy to measure with a low-power CW probe laser, fast collecting optics and a photomultiplier as shown in figure 7.2. This technique also has the advantage that it is non-invasive and can be run in parallel with a cluster-interaction experiment. Temporal resolution of the scattered light signal can also be extremely useful for setting the opening time of a pulsed gas jet and for investigating the temporal evolution of the formation of clusters. The gas stream from such a pulsed jet might take several milliseconds to build up and the pulse of heat generated in a cooled jet by operation of the solenoid can actually turn off the formation of clusters on a similar timescale [12].

The threshold for observing optical scattering from a cluster jet typically occurs for $N_c \approx 100$ [24] and the uncertainty in this value (of order a factor of two) is one of the key disadvantages of this technique. The pressure scaling of scattering after this point allows us to estimate the average size of clusters (provided that the mean density as a function of pressure is also well known). However, it is important to note that this method tells us very little about the cluster-size distribution [28].

Since most atoms in the jet condense into clusters for $\Gamma^* > 100$, it

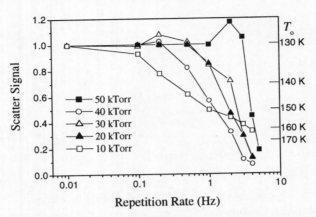

Fig. 7.4. The peak 90°-scattering signal (normalised to 1 for low repetition rates) for Ar at 10, 20, 30, 40 and 50 kTorr at a temperature of 130 K as a function of the jet-repetition rate, showing a fall off in cluster signal both for higher repetition rates and for lower backing pressures. The right-hand axis gives T_0 estimated from a scaling $N_c \propto T_0^{-5}$, from [12].

follows that the number density n_c of clusters will be given by the density of monomers before clustering, n_0 divided by N_c; hence $n_c \approx n_0/N_c$. For spherical clusters we have the radius $R_c \propto N_c^{1/3}$ and the scattering signal S_{RS} is proportional to $n_0 N_c$. We know that the density of monomers before clustering [12] is typically proportional to the backing pressure P_0 and so have $S_{RS} \approx P_0 N_c$. Finally, the Hagena parameter shows that $N_c \propto P_0^{2.0-2.5}$ and so the scattered-light signal S_{RS} should vary as $P_0^{3.0-3.5}$. Experimentally we find [12] that the scattered signal does typically vary as $P_0^{3.0-3.5}$ (see figure 7.3(a) for example) for a wide range of conditions. The average size of clusters at a particular backing pressure can then be estimated from scattering data of the type shown in figure 7.3(a). Here we assume that the observed onset of optical scattering corresponds to $N_c \approx 100$ [24] and use the measured increase in scattered signal to calculate the size of clusters as a function of the backing pressure.

We can examine the scaling of the formation of clusters with the valve temperature in a similar way. Clustering of light gases such as H_2 and D_2 (with low k values, table 7.1) that are of interest for thermonuclear-fusion experiments requires operation of jets at low temperatures in order to achieve significant formation of clusters. This can be done by cooling a jet with a suitable cryogen (liquid N_2 or He for example), either directly [32] or with a secondary heat exchanger [12] for continuous control of temperature. Figure 7.3(b) shows an optical scattering measurement of the size of D_2 clusters as a function of temperature from a He-cooled pulsed gas jet [12]. Consideration of the variations of the sizes of clusters,

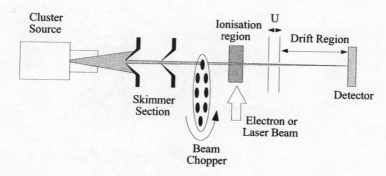

Fig. 7.5. A simple schematic diagram of a time-of-flight technique for measuring cluster-size distributions.

average gas density and Rayleigh cross section as functions of temperature suggests that the scattered signal S_{RS} should scale as T_0^{-5}, which is found to fit the measured data well. Measurements of this kind can also be used 'in reverse' to estimate the initial temperature T_0 in situations in which it might be difficult to measure it directly. A particular example is the operation of cryogenically cooled gas jets at high repetition rates. Ohmic heating of the jet assembly by the large current pulse required to drive a solenoid valve can result in quite significant rises in temperature and a consequent fall off in size of clusters (see figure 7.4 for example).

Time-of-flight spectroscopy

Time-of-flight (TOF) spectroscopy has been used in conjunction with a method of charging and accelerating clusters to determine cluster-size distributions [19, 33, 34]. Clusters in a diffuse beam defined by a number of skimmers are weakly charged using attachment of electrons, impact by electrons or photo-ionisation such that the Coulomb forces are insufficient to break the cluster apart. They can then be accelerated in a potential U and the time taken to reach a detector through a field-free drift region measured (see figure 7.5).

The time of flight of clusters is proportional to $U(m/q)^{1/2}$, where m is the mass and q the total charge of the cluster. Ideally, clusters with identical mass-to-charge ratios will have equal times of flight through the device. However, the initial velocity distribution of the clusters produced by the jet and the finite size of the ionisation region will limit the resolution of this technique. In addition a range of overlapping mass distributions will typically be recorded for clusters with different charge states and there may be cluster-size dependences in rates of ionisation, leading to skewing of measured distributions in favour of particular sizes of cluster.

As well as being a useful method of determining cluster-size distribu-

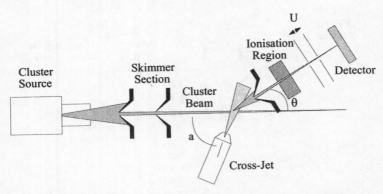

Fig. 7.6. A simple schematic diagram of a crossed-jet deflection technique for measuring cluster-size distributions.

tions, this technique can also be used for selecting the size of clusters for experiments [35, 36]. By placing an interaction experiment in the drift region (downstream of the acceleration potential), clusters with a limited range of sizes can be selected if the interaction takes place at a well-defined time after the beam of clusters has been pulse accelerated. Varying the time delay will change the range of sizes sampled. The disadvantage of this technique is that the skimming, chopping and dispersal of the beam results in extremely low densities of clusters in the interaction region. It does, however, allow very careful measurements of small (few-atom) clusters to be made with resolution of a single atom per cluster [37, 38].

Crossed-jet deflection mass spectrometry of clusters

The TOF technique outlined above requires a method of weakly ionising clusters without also destroying them. For the technique to work well the size dependence of the ionisation mechanism needs to be fully understood. However, there is another (rather technically demanding) method for characterising cluster sources that works with neutral species and so avoids this problem. Crossed-jet deflection [39, 40] (figure 7.6) utilises transfer of momentum between a beam of neutral clusters and a stream of a neutral atomic or molecular gas (containing no clusters) to produce a cluster-size-dependent angular distribution. This distribution can be found by subsequently ionising, accelerating and detecting clusters that emerge at specific angles to the original direction of the beam.

If the cluster source produces a narrow initial velocity distribution for each size of cluster and densities are sufficiently low that only single collisions between clusters and atoms from the crossed jet occur, a deflection angle can be mapped to a specific size of cluster if collisions are completely inelastic. Simple consideration of conservation of momentum gives a relationship among the angle of deflection θ, the crossed-jet angle a and the masses of the cluster and deflecting atom or molecule:

$$\tan \theta = \frac{\sin a}{\cos a + \frac{m_1}{m_2}\frac{u_1}{u_2}}, \qquad (7.4)$$

which for $m_1 \gg m_2$ reduces to

$$\theta = \frac{m_2 u_2}{m_1 u_1} \sin a, \qquad (7.5)$$

where m_1 and u_1 are the pre-collision mass and speed of the clusters and m_2 and u_2 are the pre-collision mass and speed of the crossed-beam particles. This angular distribution needs to be convoluted with a scattering-probability function as described in [41] to find the actual cluster-mass distribution. Both the TOF technique and the crossed-jet technique are sufficiently complex and demanding that they are not really appropriate as on-line diagnostics of a dense-cluster experiment carried out near the nozzle of a gas jet.

7.5 Measurements of the cluster-gas density and jet profiles

Pulsed gas jets can currently provide high average densities (up to 5×10^{19} atoms cm^{-3}) [12] close to the nozzle over volumes of a few mm^3. These bulk cluster media are of particular interest in the investigation of effects with strong particle-density scalings. These include thermonuclear fusion and generation of harmonics, both of which should scale as the number density n^2. Clearly a method of determining the absolute cluster density and the spatial profile of the cluster jet is important in investigations of this type.

Near the nozzle the plume from a gas jet is typically quite well colli-mated (the exact degree of collimation will depend on the profile of the nozzle and whether it is sonic or supersonic) and it can be shown that, to first order, the density, $N(z)$, of gas on the axis (z) of a sonic expansion will vary approximately as [42]

$$N(z) = \frac{N_0}{1 + \alpha z^2}, \qquad (7.6)$$

where N_0 is the density of gas near the nozzle and α is a parameter that depends on the geometry and backing pressure of the jet. However, we have ignored the non-uniform radial density profile and its effect on the axial density profile in (7.6). We can also use the more rigorous treatment of the expansion of a gas jet into vacuum outlined in [20], in which the details of the radial profile are considered. Free flow of a low-viscosity fluid into vacuum often results in self-similar exponential axial density profiles and Gaussian radial density profiles [43, 44]. For a low-viscosity fluid of spatially constant temperature, simple solutions of the Navier–Stokes equation describing the fluid's momentum balance produce these density profiles. For such a fluid, the Navier–Stokes equation can be written as

$$mN(\boldsymbol{u}\cdot\nabla)\,\boldsymbol{u} = -\nabla p = -k_\mathrm{B}T\,\nabla N \tag{7.7}$$

for a spatially constant temperature gas. Here $\boldsymbol{u}$ is the velocity of the fluid, p is the pressure of the fluid and m is the mass of the gas molecule or atom. If we assume cylindrical symmetry and equate the radial and axial components of equation (7.7), we can write

$$mNu_r\,\frac{\partial u_r}{\partial r} = -k_\mathrm{B}T\,\frac{\partial N}{\partial r}, \tag{7.8}$$

$$mNu_z\,\frac{\partial u_z}{\partial z} = -k_\mathrm{B}T\,\frac{\partial N}{\partial z}, \tag{7.9}$$

where u_r and u_z are the radial and axial components of the fluid's velocity. Equations (7.8) and (7.9) assume that there is no azimuthal (θ) fluid-velocity component and that the change in u_r with the axial distance z is much less than the change in u_r with the radius r (which is generally the case for expansion of a gas jet from a nozzle of aperture <1 mm). If we assume that the radial acceleration is such that the radial component of velocity u_r varies linearly with the radius r, namely $u_r = \kappa r$, where κ is constant with respect to r, then we can solve equation (7.8) to give a Gaussian radial density profile:

$$N(r, z) = N(r, 0) \exp\!\left(\frac{-m\kappa^2 r^2}{2k_\mathrm{B}T}\right). \tag{7.10}$$

If we further assume that the axial acceleration $u_z\,\partial u_z/\partial z = a_0$ is constant with respect to z then equation (7.10) can be solved to give an exponential axial density profile:

$$N(r, z) = N(r, 0) \exp\!\left(\frac{-ma_0 z}{2k_\mathrm{B}T}\right). \tag{7.11}$$

A simple modification of the Rayleigh-scattering technique described in section 7.4 can be used to investigate cluster-jet profiles. Replacing the PMT in figure 7.2 by a CCD imaging system at 90° to the probe laser and imaging the side scattering will give a rough indication of the cluster-jet profile. Care needs to be taken in interpreting results from this method, however, as the average size of clusters (and hence the Rayleigh-scattering cross section) will vary across the jet's profile.

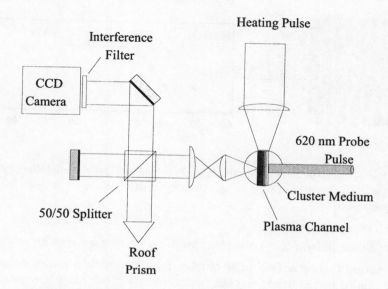

Fig. 7.7. A schematic diagram of a Michelson-type interferometer used for measuring the absolute electron density in an ionised cluster medium.

Interferometric profiling of cold gas jets

In principle optical interferometry can be used as a less error-prone method of measuring gas-jet profiles than imaging of Rayleigh-scattered light. Unfortunately the refractive index of a cluster or monatomic gas medium produced by a pulsed jet is typically very close to unity. This makes direct interferometric profiling difficult, but it can be achieved with a sufficiently sensitive technique [20, 21]. There are also uncertainties in the refractive index of a cluster medium compared with a monatomic gas of the same average density.

In general the number density N_0 of molecules in a gas can be deduced from the refractive index n [45], by using

$$N_0 = \frac{n^2 - 1}{3Am_{\mathrm{H}}} \approx \frac{n - 1}{3Am_{\mathrm{H}}}, \tag{7.12}$$

where A is the molar refractivity and m_{H} is the mass of the hydrogen atom. For visible wavelengths, molar refractivities (see for example the CRC handbook) are constant with density and temperature near room temperature.

It should be noted that a cluster medium is composed of microscopic particles with near-solid or near-liquid densities and so use of a refractive index for a monatomic gas of the same average density might result in some errors.

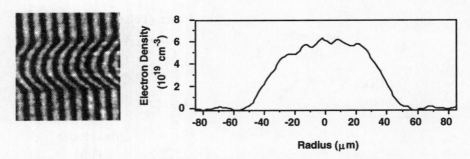

Fig. 7.8.   A short-pulse interferogram and the deconvoluted radial density profile of a 2-ps, 1053-nm laser pulse at an intensity of 10^{16} W cm^{-2} focused into a jet of D_2 gas, from [46].

Short-pulse interferometric profiling of ionised cluster jets

The energetic interaction of a cluster medium with a short-pulse laser allows one to use a rather elegant technique for measuring absolute densities and density profiles of cluster jets [46]. A short-pulse, high-intensity laser can be used to heat and ionise a bulk cluster medium efficiently. This liberates free electrons and the change in refractive index that results is particularly amenable to interferometric measurements. Use of a picosecond probe laser can freeze out hydrodynamic motion of the cluster plasma and this technique has the additional advantage that it provides absolute measurements of density and density profiles without reference to a test cell.

Figure 7.7 shows an experimental arrangement for measurements of this kind, which were carried out at Imperial College. An intense pulse is focused into a gas jet, causing rapid and complete ionisation in the focal volume. A probe pulse then illuminates this ionised region at 90° to the axis of the heating pulse about 10 ps after its arrival. An interferogram of the probe is recorded using a Michelson arrangement, then Abel inverted to recover phase information, whereupon an electron-density profile can be calculated. Figure 7.8 shows a typical interferogram and deconvoluted density profile from a picosecond 1054-nm pulse focused at 10^{16} W cm^{-2} into a jet of D_2 clusters. The measured electron density for D_2 at 293 and 79 K scales linearly as a function of the backing pressure in this experiment [12]. The number of electrons liberated per molecule is well known and the absolute atomic density can be obtained as a result. In [46], peak atomic densities of 4×10^{19} atoms cm^{-3} were found for a pulsed solenoid gas jet of the type shown in figure 7.1 backed with 50 kTorr of D_2.

An important advantage of this method is that, due to the short timescales involved, no significant hydrodynamic motion occurs during the

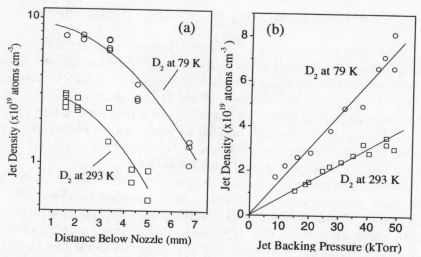

Fig. 7.9. (a) Average atomic density produced by a solenoid pulsed gas jet at 103 K and 290 K as a function of distance below the nozzle [46]. (b) The same as (a) as a function of jet backing pressure [46].

measurement. It also removes some uncertainties concerning the refractive index of clustered media in the interferometry of cold jets of clusters of the type outlined in section 7.5. Varying the position of the heating beam also allows the full density profile of the jet as a function of distance from the nozzle to be found. Figure 7.9(a) shows the measured atomic density as a function of distance below the nozzle [46] for D_2 at 79 and 293 K with fits to Gaussian profiles. The approximate $T_0^{0.5}$ scaling between the two plots is accounted for by the additional gas released when the valve is cooled. However, there is little change in the profile of the jet when the valve is cooled and so the interaction length for an intense laser remains essentially unchanged. Figure 7.9(b) shows the linear variation in average atomic density with the backing pressure of the jet determined using the same technique.

By holding $P_0 T_0^{-0.5}$ constant when changing the size of the clusters, the mean atomic density can also be held nearly constant. This makes accounting for density-dependent effects in the interaction simpler and, in high-intensity laser-interaction experiments, avoids the unexpected occurrence of processes such as stimulated Brillouin scattering that grow non-linearly with density. However, in experiments in which the size of cluster is varied by controlling the temperature it is often more convenient to work at a constant backing pressure and account for the change in density in the analysis, provided that the physics of the interaction is sufficiently well understood.

7.6 Laser absorption in extended cluster media

The extremely strong coupling of single clusters to an intense laser field suggests that an extended cluster medium should be an efficient absorber of laser pulses shorter than the $\simeq$100-fs timescale on which clusters explode. As we shall see in this section, bulk cluster media are indeed extremely efficient at absorbing intense laser pulses and in fact even relatively long pulses (2 ps [47] and 30 ps [48]) can be absorbed in these media, in part due the modification of the temporal profile of a 'long' laser pulse propagating through such a medium.

Studies of interactions of laser beams with solid targets have shown that such plasmas typically absorb a large fraction (between 10% and 80%) of the incident energy, depending upon the intensity and wavelength. Experiments of this type show that a large amount of energy can be deposited per unit area and that high temperatures (>100 eV) are achievable [49, 50]. However, in experiments on solid targets the extremely rapid conduction of heat into the cold substrate beneath the plasma typically clamps the plasma temperature to <1000 eV [50, 51]. In addition most of the laser energy deposited in solid targets is coupled to the plasma electrons, which typically cool too rapidly to transfer much of this energy to the ions. Cluster media are of particular interest for creating hot-ion plasmas, since there is limited loss of energy by conduction of heat and the explosion of clusters efficiently transfers energy from electrons to the cluster ions.

The first absolute measurements of the fractional absorption of intense, short-duration laser pulses in extended cluster media were carried out at Imperial College [47]. The absorption of 2-ps pulses from an Nd:glass CPA laser system (frequency doubled to 527 nm to eliminate any low-intensity pre-pulse) by a range of clustered and monatomic gases (both high and low Z) was measured as a function of the vacuum laser intensity. Pulses of energy up to 0.5 J were focused with an $f/12$ lens to a spot of $1/e^2$ diameter 20 μm onto the output of a pulsed gas jet [12]. The laser focus was placed approximately 2 mm below the jet's nozzle, giving an interaction region about 2 mm long, containing clusters or a monatomic gas with a mean average density up to 5×10^{19} atoms cm^{-3}.

The laser light transmitted through the cluster medium was collected with an 8-cm diameter $f/2$ lens to ensure that any light refracted by formation of a plasma at the focus was sampled. The energy of the transmitted light was measured with a volume-absorbing calorimeter. The input laser energy was monitored with a photodiode and backscattered light was also monitored (although none was observed). Imaging from the side of the plasma was also performed and showed that no significant amount of laser energy was side scattered (<1 mJ). Figure 7.10 summarises results

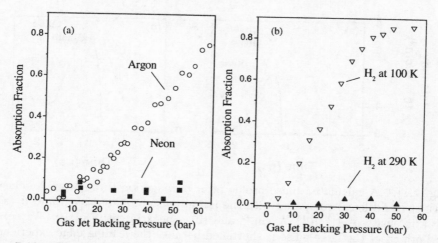

Fig. 7.10. The fraction of laser light absorbed in a gas jet as a function of the backing pressure with a peak laser intensity of 7×10^{16} W cm^{-2}. (Each point is an average of three laser shots.) (a) Argon and neon gas jets. (b) Room-temperature and cryogenically cooled jets of H$_2$ gas. Taken from [12].

from this series of experiments. Figure 7.10(a) shows that the absorption of Ne (which, under the conditions in this experiment, is monatomic) is extremely low (of order a few per cent). However, a medium of Ar clusters exhibits increasingly strong absorption as the backing pressure of the gas jet (and hence the size and average density of clusters) is increased. At 60 bars, when the medium is composed of $\simeq$80-Å clusters of Ar with a mean density of 2×10^{19} atoms cm^{-3}, nearly 80% of the incident light is coupled into the medium. Similar results are found for higher-Z materials such as Xe and, even more surprisingly, for low-Z materials as well. Figure 7.10(b) shows the absorption as a function of the backing pressure for H$_2$ at 290 K (at which there is very little formation of clusters) and H$_2$ at 100 K (at which there is strong clustering). This result is particularly intriguing insofar as the ability to create high-temperature hydrogen plasmas efficiently points the way towards possible applications in thermonuclear fusion.

The results shown in figure 7.10 clearly indicate that intense laser pulses can couple efficiently to bulk cluster media, even when the duration of the pulse is relatively long compared with the timescale for ionisation and explosion of an individual cluster. This should not be too surprising since in an extended, highly absorbing medium, the leading edge of a laser pulse can be rapidly eaten away. The risetime of the pulse becomes steeper as a result of preferential absorption at its leading edge, so absorption becomes even more efficient. The interaction of an energetic laser pulse with an extended cluster medium is thus highly dynamic, with different amounts

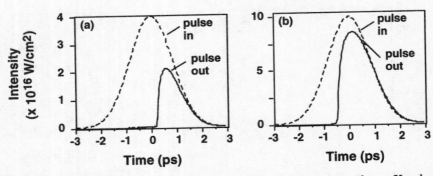

Fig. 7.11. A calculated pulse profile after propagation through an Xe-cluster gas jet (solid lines) compared with the initial pulse profile (dashed lines): (a) a peak intensity of 4×10^{16} W cm^{-2} and (b) a peak intensity of 10^{17} W cm^{-2}. In both cases the laser pulse is shortened because the leading edge experiences preferential absorption.

of energy being deposited at various points along the axis of propagation of the laser beam.

To investigate some of these effects, numerical modelling of the propagation of a relatively long (2 ps) laser pulse into an extended Xe-cluster medium [47] has been carried out using a plasma model of heating of clusters detailed in [52]. This model calculated the energy deposited in an expanding cluster microplasma, accounting for inverse *Bremsstrahlung* heating of electrons, tunnel and collisional ionisation and contributions from both hydrodynamic and Coulomb terms to the expansion of clusters. The effect of multi-photon ionisation at low intensities (which seeds the initial avalanche breakdown of the cluster) was included by incorporating ionisation cross sections from [53]. This model was solved for a pulse propagating through a series of spatial cells, in which the absorption dynamics of clusters was found using realistic density profiles of the type described in section 7.5. The total absorption was calculated as a function of intensity and integrated over a Gaussian focal-spot intensity distribution (ignoring any refractive effects). The rapid expansion of the laser-heated clusters causes them to disassemble on a timescale of 100–200 fs, significantly faster than the 2-ps pulse modelled here. As a result, the leading edge of the pulse experiences the highest absorption as it propagates into the cluster medium. This leads to the effective pulse shortening shown in figure 7.11.

This result also suggests that, for sufficiently energetic laser pulses, the cluster medium can 'burn through' and the fractional absorption should then start to fall again. This effect is elegantly demonstrated in figure 7.12. Here the absorption of 2-ps, 526-nm laser pulses measured in an extended medium of $\simeq$100-Å clusters of Xe is compared with numerical

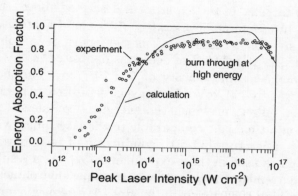

Fig. 7.12. The fraction of the laser beam's energy absorbed in a jet of Xe clusters of size $\simeq$100 Å as a function of the peak intensity, from [47]. Each point is a single laser shot and the solid line is a calculation using the 'nanoplasma' model of absorption.

modelling results.

Above 2×10^{12} W cm^{-2} the absorption increases rapidly as the laser pulse becomes sufficiently intense to generate free electrons in the clusters and the nanoplasma mechanism of heating turns on. Between 10^{14} and 5×10^{16} W cm^{-2} the absorption is extremely high (nearly 90%) and above 5×10^{16} W cm^{-2} the fractional absorption falls off as the pulse burns through the cluster medium. The model provides a remarkably good fit to the data, particularly above 10^{14} W cm^{-2}. Below this intensity the model underestimates the measured absorption fraction and it seems likely that this is the result of low-probability multi-photon processes giving rise to a few free electrons per 1000-atom Xe cluster, which then seed a rapid avalanche breakdown at intensities lower than expected.

The modelling indicates that the 'burn-through' feature in the data is the result of the limited time during which the clusters absorb laser energy. Once the peak intensity is sufficiently high, the pulse 'burns through' the cluster medium and the trailing edge of the pulse propagates without significant absorption. An interesting point to note is the very substantial difference between the absorption of a cluster medium and that of a monatomic gas or the ionised medium produced when clusters are destroyed by a laser pulse. This suggests ways in which a cluster medium can be 'machined' to give absorption and deposition of energy in specific spatial regions. A number of applications, for instance in the guiding of intense pulses, could be brought about in this way, and this technique is discussed in more detail in section 7.7.

The very high absorption observed in these measurements indicates that a large amount of laser energy can be deposited in a small volume in these cluster targets. In the experiment described above, the laser energy is deposited in a volume of $\simeq \pi \omega_0^2 l \approx 10^{-6}$ cm^3 (where ω_0 is the radius of the focal spot of the laser and l is the length of the plasma). At the highest intensity, over 0.25 J of energy is deposited in this small volume. With an average atomic density of 5×10^{18} cm^{-3} in the gas jet, this implies that up to 300 keV of energy is deposited per atom. However, unlike a solid target, in which the energy is conducted away from the hot plasma on a timescale comparable to the laser pulse width, clamping the plasma temperature [50, 51], conduction of heat is very slow in the low-density gases used in these experiments [52]. As a result, the plasmas created are extremely hot and present a unique environment for studying and utilising high-energy-density matter. We review a number of these applications in the next few sections.

7.7 Modifying effects of propagation and deposition of energy

The extremely high efficiency of absorption of short, intense laser pulses by a cluster medium points the way to some rather elegant methods of tailoring cluster media for experiments in areas such as the guiding of intense laser pulses and shock physics under conditions that scale to astrophysically interesting regimes.

Atomic clusters are rather fragile and can be destroyed in a number of ways, including weakly heating and ionising them with a laser beam focused to an intensity well below the 10^{12} W cm^{-2} threshold for energetic cluster effects in figure 7.12. This technique has been used to carry out detailed comparisons of media that are identical in all respects save for the presence or absence of clusters in a constant-density background [54]. In chapter 6, the use of this technique to investigate the non-linear optical properties of clusters is described in more detail. A weak ultra-violet (UV) pulse at an intensity of $\simeq 10^{10}$ W cm^{-2} was employed to destroy clusters. This technique can also be applied to the control of the deposition of energy in limited volumes of a cluster medium.

In [55] the interferometric technique described in section 7.5 and [46] was used to investigate spatial tailoring of the deposition of energy in a cluster medium for the first time. Two 526-nm optical pulses (a weak pre-pulse and a strong heating pulse) were focused with an $f/12$ lens to a spot diameter ($1/e^2$ in vacuum) of 30 μm in a jet of Ar gas, which produced an average atomic density of 1.5×10^{19} cm^{-3}. Under these conditions the average cluster size is $\simeq 5000$ atoms. The time delay between the pre-pulse and heating pulse was of order 50 ps and the pre-pulse (focused to an intensity of $\simeq 10^{13}$ W cm^{-2}) resulted in a small amount of ionisation along the axis of the laser beam. This can be seen in the interferogram taken at $t = -15$ ps in figure 7.13. The deconvoluted plasma profile, shown next

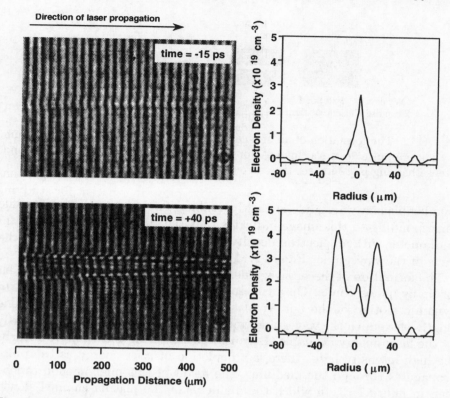

Fig. 7.13. Interferometric images of a plasma channel in an argon-cluster medium when a weak pre-pulse is used to destroy clusters along the axis of the laser beam, from [55]. The images show the gas 15 ps before the main pulse enters the medium and 40 ps after the main pulse has traversed the medium. The deconvoluted electron-density profiles are shown on the right-hand side.

to the image, indicates that the average ionisation of the channel is about 2+ (an electron density of 3×10^{19} cm^{-3}) and that all the clusters in this region would have been destroyed as a result.

The use of this small pre-pulse dramatically changes the dynamics of the formation of the plasma channel compared with the single-pulse data shown in figure 7.8. As the main heating pulse propagates into the cluster medium along the same axis as the pre-pulse, blurring of the fringes in the laser-heated region is caused by the extremely rapid ionisation of clusters and a varying electron density that changes on a timescale shorter than the $\simeq$1-ps probe pulse. However, the bulk of the ionisation occurs in a region outside the channel created by the pre-pulse because there is little additional deposition of energy in the cool Ar plasma compared with the surrounding cluster medium. At $t = +40$ ps a plasma channel has formed and deconvolution of this interferogram shows that a hollow

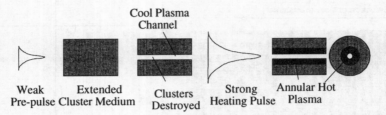

Fig. 7.14.   The formation of an annular plasma waveguide in a cluster medium. A weak pre-pulse destroys clusters along the axis of the laser beam and a second, intense heating pulse ionises the surrounding material.

annular plasma waveguide is clearly visible. The electron-density profile corresponding to this image indicates that the channel is roughly 25 μm in diameter with an electron density reaching about 4×10^{19} cm^{-3} in the walls of the waveguide. Figure 7.14 summarises this process.

The formation of these waveguides is very reproducible and they persist for many nanoseconds. On a timescale of tens of nanoseconds the hydrodynamic motion of the hot Ar plasma causes closure of the guide. The length over which the waveguide can be produced is determined primarily by the laser energy available. Because the clustering gas strongly absorbs the high-intensity light, there is a depletion of laser energy as the pulse propagates through the medium. The result of this phenomenon can be seen in figure 7.13, in which the diameter of the plasma channel slowly decreases along the axis of propagation of the laser beam. Refraction, however, will ultimately limit the maximum length achievable, though, because of the modest intensities required, quite weakly focused pulses can be used and clamping of the peak intensity due to refraction should not be a significant problem.

The formation of a plasma channel, which was demonstrated in this experiment, is significant because it provides an elegant demonstration of a more general technique. The modification of cluster density by a weak 'machining' laser can be used to impose a large modulation on the spatial deposition of energy in a cluster medium. This has potential applications in, for example, creating gas media with modulated deposition of energy that would be useful in studying the growth of Rayleigh–Taylor instabilities under conditions of interest to astrophysics, but amenable to laboratory-scale experiments. Figure 7.15, for example, shows how a modulation could be imposed along the axis of a heating laser beam by using a pair of 90° transverse 'machining' beams to write a 'grating' into the medium.

The annular density profile produced using an axial pre-pulse also has intriguing possibilities in its own right. Some important applications in high-intensity laser–matter-interaction physics require the propagation of

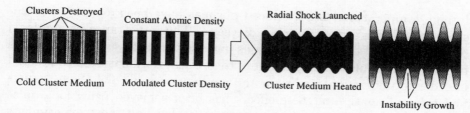

Fig. 7.15. A schematic diagram of how machining of a cluster medium could be used to prepare a modulated shock for the study of Rayleigh–Taylor instabilities in a hot cluster plasma.

a focused, ultra-short laser pulse through an ionising gas at near atmospheric density ($\simeq 2 \times 10^{19}$ cm^{-3}). These applications, which include use for X-ray lasers [56] and the acceleration of electrons [57], require plasma channels of lengths between a few millimetres and a few centimetres. The creation of such plasma channels with a focused, intense laser pulse has posed a persistent problem because the radial refraction of a plasma created by ionisation acts as a negative lens and defocuses the laser beam [58]. This effect tends to limit the maximum focused intensity that can be achieved in a gaseous medium and shortens the effective length over which an intense pulse can propagate [59]. A significant amount of effort has been devoted to the study of relativistic [60] and charge-displacement [61] self-channelling of intense pulses in under-dense plasmas. However, these phenomena require very high laser intensities (usually $>10^{18}$ W cm^{-2}).

A channel of the type shown in figure 7.13 has a number of advantages over the techniques described in [60, 61]. If the pre-pulse is of short duration and the delay between the pre-pulse and the channel-forming pulse is less than the time required for hydrodynamic expansion of the pre-pulse channel (of the order of a few nanoseconds), the average density of ions within the waveguide will not be reduced below the initial atomic density. Only the charge state of the ions and the electron density are lower than those of the surrounding medium. This is in contrast to the expansion of a pre-heated plasma, as was demonstrated in [62], in which the waveguide was formed by the rarefaction of ions in the centre of the channel through hydrodynamic expansion. This drop in density is undesirable if a high density is required in the centre of the channel.

7.8 The transport of energy in dense cluster media

Rapid local heating of part of a cluster medium to generate a high-temperature, highly ionised plasma or a pulsed source of super-thermal or 'hot' electrons provides a route to studying non-local transport of energy and the formation of shock waves. The mechanisms for transport of heat in high-temperature laser-produced plasmas have been a topic of extensive research for a good number of years. The importance of non-local transport of heat due to hot electrons in these plasmas was identified some time ago and a substantial amount of theoretical work has been conducted in order to understand its origin and consequences [63–66]. Some experimental confirmation of these effects has been achieved using long-pulse (>500 ps) irradiation of solid targets [67], although most experiments have been limited to the observation of indirect effects on processes such as mass ablation [68]. Interpretation of these measurements is further complicated when long-pulse lasers are used since energy is deposited in the plasma on the same timescale as the heat-transport dynamics. Intense, ultra-short (picosecond or femtosecond) laser pulses can, however, be used to make clean, time-resolved studies of heat-transport phenomena [69] and are ideal for more direct measurement of non-local transport effects.

Temporally and spatially resolved measurements of the electron-density profiles of the hot plasma filament using short-pulse laser interferometry described in [47] and section 7.5 have been used to study non-local transport in clustered gases [70]. Measurements of a rapid heat-transport-driven radial ionisation wave were made on a timescale much faster than the hydrodynamic expansion of such plasmas [71] and it was found that this transport of energy was dominated by strong non-local effects. Through comparison with numerical modelling these measurements provide a direct experimental comparison with the theory of Luciani *et al.* [72, 73] on non-local transport of electrons.

In plasmas with modest temperature gradients the flow of energy is usually dominated by diffusive transport of heat by electrons. In this regime the thermal conductivity is given by the usual Spitzer formula [74]. However, when the heat gradients in the plasma are large, the diffusive approximation breaks down. This failure occurs when the mean free path of the electrons approaches the spatial scale of the temperature gradient. If the mean free path, λ_e, is given by the usual formula for 90° scattering [75], non-local transport dominates when [72]

$$\frac{\lambda_e}{T_e}\frac{\partial T_e}{\partial x} = \frac{T_e}{4\pi e^4 n_e (Z+1)^{1/2}\ln\Lambda}\frac{\partial T_e}{\partial x} \geq 2\times 10^{-3}, \qquad (7.13)$$

where T_e is the electron temperature, n_e the electron density, e the electron's charge, Z the ion charge state and $\ln\Lambda$ the Coulomb logarithm. In this regime the transport of heat is dominated by hot electrons free-streaming from the hot portions of the plasma to the cooler regions and the Spitzer–Härm formula for the conductivity exceeds the maximum possible rate of conduction by free-streaming electrons (given as $q_{fs} = 3n_e T_e^{3/2}/(2m_e^{1/2})$), where m_e is the mass of the electron. This regime

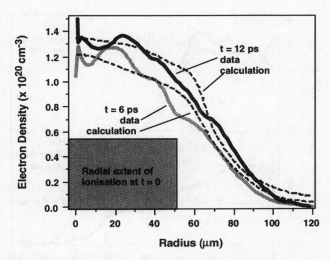

Fig. 7.16. Deconvoluted electron-density profiles measured in an argon-cluster plasma at two times (solid lines), taken from [70]. The dashed lines are the results of Med103 hydro-code calculations including the effects of non-local transport of heat by electrons.

is usually modelled by clamping the heat conductivity to a value given by fq_{fs} where f is an empirical constant less than unity, called the flux-limiter [75]. To model the effects of this regime of heat transport more accurately, a semi-empirical formula for the non-local transport of energy was proposed by Luciani *et al.* [72]. Results of calculations using this formula have been shown to be in good agreement with more detailed Fokker–Planck simulations of transport of energy by electrons in steep gradients and deviate substantially from simple flux-limited calculations [64, 73].

In the work described in [70] an extended cluster medium was produced with a pulsed jet of Ar gas, which created a gas with an average atomic density of up to 1.5×10^{19} cm^{-3} composed of clusters with an average size of $\simeq 5000$ atoms. The cluster gas was then heated in a small region with a 2-ps, 526-nm laser pulse focused to a $1/e^2$ diameter of 50 μm. Under these conditions roughly 60% of the total incident laser energy of 300 mJ was absorbed by the gas within the focal volume. For argon with an average ionisation state of 8+ (Ne-like), this fraction implies an initial electron temperature of $\simeq 1.5$ keV, assuming that one has equal electron and ion temperatures. The rapid transport of energy from the laser-heated region of the cluster medium to the surrounding cold cluster gas (which then ionised) was followed with a picosecond optical probe.

Deconvoluted electron-density profiles from argon-cluster plasmas at two times are shown in figure 7.16. Since the initial radius of the ionised

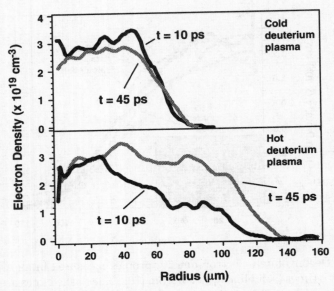

Fig. 7.17. Measured electron-density profiles in deuterium gas at two times, from [70]. The top plot is for a cold plasma that contained no clusters. The bottom plot is for a plasma created in a gas containing D_2 clusters, in which the absorption of laser light and initial plasma temperature were high.

region was $\simeq 50$ μm, this plot shows that the velocity of the ionisation wave driven by hot electrons is of order 10^7 m/s. This is roughly consistent with the velocity of free-streaming thermal electrons in a 1.5 keV-plasma $(1.6 \times 10^7$ m s$^{-1})$.

A graphic illustration of the importance of hot electrons in plasma transport dynamics in a bulk-cluster medium is given by the plots in figure 7.17, which shows the plasma profiles at two times for a deuterium gas with and without clusters. In the first case the plasma is created in a gas of D_2 molecules from a gas jet at room temperature, a case in which the absorption of laser light is very low ($<3\%$) and the plasma's temperature is of order 20–30 eV. In the second case, the evolution of a plasma created in deuterium clusters when the gas jet is cooled to -170 °C is illustrated. Here the absorption is large ($>80\%$) and the plasma is much hotter. The low-temperature deuterium plasma exhibits a flat-topped profile, a result of complete ionisation of the gas by the laser at $t = 0$, with no evidence of a fast radial heat wave at any later time. The high-temperature plasma, however, exhibits a fast ionisation wave travelling outwards from the initial plasma in a manner similar to that seen in the argon plasma.

Results of the type shown in figures 7.16 and 7.17 have been compared with detailed hydrodynamic simulations carried out with the MED103 code [77]. Although these simulations include all of the equations for hydrodynamic motion of the plasma, it is worth noting that virtually no such motion occurs on the timescales considered here as a result of the short duration of the heating pulse.

Ionisation and excitation of the plasma in this model are governed by time-dependent non-local-thermal-equilibrium population-rate equations and the formulation of non-local heat transport by Luciani, Mora and Virmont was included to test its accuracy using a variation of their original formula to account for cylindrical geometry [76]. It was found that the non-local heat-transport model predicted the shape and extent of the radial ionisation profile quite accurately (though there is a small discrepancy at large radii, possibly due to an inaccurate estimation of the mean free path of electrons in the cold argon gas).

As a more general comment it is worth noting that the comparison of modelling and measurements of heat flow in cluster media made in [70] and described above indicates that non-local transport of heat by electrons is an important mechanism in bulk cluster plasmas due to the very high electron temperatures reached and the large thermal gradients generated. A more conventional treatment of flux-limited diffusive heat transport is inadequate for predicting the heat flux in such a medium and will thus be inaccurate when predicting the spatial extent of the emission of X-rays from the plasma.

7.9 The propagation of shocks in cluster media

The high temperatures and strong plasma gradients described in the work above can be used to investigate a number of important plasma processes in the laboratory. In particular, bulk cluster media have been used in recent experiments at the LLNL [78, 79] to investigate the propagation of shocks in regimes of great interest to astrophysics.

Understanding the dynamics of shock waves is crucial to understanding the structure of the interstellar medium [80, 81]. The evolution of shock waves and the growth of hydrodynamic instabilities such as the Rayleigh–Taylor and Richtmyer–Meshkov instabilities play fundamental roles in the expansion of blast waves from supernovae and their interaction with the diffuse interstellar medium [82]. In such interstellar shocks, the effects of radiation transport are important and can dramatically affect the shock dynamics by radiative cooling and via the formation of a radiative precursor ahead of the shock. Because of the importance of these processes in astrophysical structures, there is a strong motivation for studying them in laboratory-produced plasmas that can be scaled to astrophysically relevant conditions. Indeed, there has recently been some success regarding producing conditions of relevance to astrophysics using shocks driven by

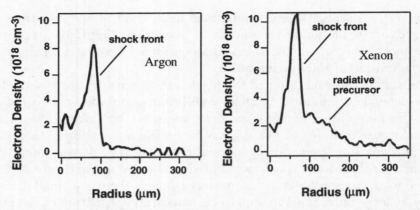

Fig. 7.18. Electron-density profiles of blast waves produced in argon and xenon gas 6 ns after heating by a 30-fs, 17-mJ laser pulse, from [78]. The argon gas was produced with a gas pressure of 60 bars backing a sonic gas nozzle (with a 0.5-mm orifice); the xenon gas was produced with a backing pressure of 20 bars to give the same mass density.

large-scale, long-pulse lasers [83].

Long-pulse experiments of the type described in [83] are unable to provide measurements of the complete shape and temporal evolution of a radiative shock. However, the combination of high (a few kilo-electron-volts) plasma temperatures and fast timescales accessible in bulk cluster media heated with sub-picosecond-laser pulses has recently allowed this process to be investigated [78, 79]. Under these conditions a strong, cylindrically symmetric blast wave should develop and propagate out into the cold material surrounding the laser focus and this is particularly amenable to short-pulse optical probing using techniques described in section 7.5. The experiments required to investigate these shock phenomena are very similar to the work described in section 7.8 and [47], but the evolution of the ionisation front is followed on the 1–100-ns timescale rather than over a few tens of picoseconds to allow the development of hydrodynamic processes.

Figure 7.18 shows heat fronts produced in argon and xenon clustered gases, 6 ns after heating by a 30-fs, 17-mJ laser pulse. The initial laser-heated plasma filament had a diameter of 80 μm (set by the size of the laser focus) and a length of ≃2 mm (limited by the spatial extent of the cluster medium) and the jet conditions were set to give the same mass density for the Ar and Xe gases. The fraction of the total laser energy absorbed in the argon-gas jet was ≃35% whereas that for the xenon-gas jet was nearly 65%. This corresponds to deposition of 30 mJ cm^{-1} in argon and 55 mJ cm^{-1} in xenon.

A strong blast wave is seen in both cases and the velocity of the ionisation front in argon is $\simeq 2 \times 10^6$ cm s^{-1}, corresponding to a Mach-60 shock. The position of the blast-wave front also exhibits the $t^{1/2}$ time dependence expected from the self-similar solution of the cylindrical Sedov blast-wave equation [84]. The profile of the argon plasma is that of a classical blast wave with a sharp shock front. However, the blast wave in xenon exhibits a shock with substantial ionisation ahead of the front. The extent of this ionisation front decreases with time such that it is virtually unobservable after 12 ns. This precursor ionisation is attributed to the occurrence of substantial radiative transport in the xenon plasma [78, 79]. The strong UV and XUV emission of the hot xenon plasma heats the cold, ambient Xe gas prior to the arrival of the blast wave.

Experiments of this kind confirm that the plasmas created in bulk-cluster media provide access to a regime in which radiation transport becomes important in the evolution of the shock profile and work on how to use such plasmas for more studies of shock dynamics that can be scaled to areas of importance for astrophysics is in progress. The relatively long (10–100 ns) timescales required for the development of hydrodynamic instabilities in these plasmas can be addressed by using modulated deposition of energy in the shocked medium of the type suggested in figure 7.15. This would allow studies of the growth of Rayleigh–Taylor instabilities at specific wavelengths imprinted on the initial heat front.

7.10 The production of X-rays and XUV from cluster gases

The efficient coupling of laser light into cluster media and the high charge states and high energies of ions observed experimentally by a number of research groups all suggest that dense targets composed of atomic clusters will be extremely efficient emitters of X-ray radiation. Indeed, some of the pioneering experiments that pointed to there being new and exciting high-field effects to be studied in atomic clusters found anomalously energetic X-rays [85–87] being emitted from cluster targets. The lack of debris from a cluster target compared with that from a solid also points the way to 'clean' sources of soft X-rays for lithographic applications, whilst the high local density that exists within laser-heated clusters before they explode suggests that they can be used to investigate the physics of high-density plasmas through spectroscopic means. With solid targets such observations are often complicated by the fact that, although short-pulse-laser irradiation of a solid generates high-temperature, near-solid-density plasmas, radiation from these plasmas must then travel through a significant amount of hot, ionised material before reaching a detector. Absorption and re-emission of radiation in such a medium involve a number of complex processes, which can make interpretation of such spectroscopic results difficult.

As noted in chapter 5, early work by Rhodes, Boyer and others [85–87] found strong M- and L-shell emission from Kr- and Xe-cluster targets, including a component attributed to Kr^{9+}. These results are significantly more energetic than would be expected from mechanisms such as field ionisation under the conditions used. The mechanism proposed for this observation by Rhodes and co-workers was the production of inner-shell vacancies in atoms within the clusters as a result of multi-photon ionisation and they went on to suggest that such inner-shell-ionised cluster media could be used for X-ray lasers [87] since this mechanism would generate a transient population inversion. Such an inner-shell-ionisation mechanism should lead to prompt emission of X-rays because these vacancies are rapidly filled by electrons from higher-lying levels in the same atom and it has also been suggested by this group [86, 87] that the yield of X-rays should be strongly dependent on the wavelength of the pump laser, with shorter wavelengths giving much higher yields.

Both the short duration and the strong wavelength scaling of cluster X-rays predicted by the model of cluster heating in [86, 87] are in disagreement with a number of other experimental observations and, indeed, the detailed single-cluster work described in chapters 5 and 6. Early work by Ditmire *et al.* [88] used an X-ray streak camera to show that emission from laser-heated cluster plasmas was long lived and characterised by nanosecond timescales. More recent work in Lund [89] aimed at characterising cluster X-ray sources for applications has confirmed this observation.

As an aside, the X-ray work described in [88] is particularly instructive in showing how the high-field community's interest in cluster targets grew out of an 'accidental' observation. An anomalous emission line of He was found in an X-ray-laser experiment that was aimed at improving upon a short-pump-driven scheme that had first been demonstrated by Lemoff *et al.* [90]. The Lemoff scheme used a static filling of Xe held in a cell, tunnel-ionised to Xe^{9+} by circularly polarised light and operated at 418 Å. The authors of [88] were using a high-pressure linear gas jet rather than a static cell, which would in fact have prevented their X-ray laser from working because it would have caused overheating of electrons through strong coupling to Xe clusters produced in the jet. However, as part of the set-up for this experiment, generation of harmonics from a jet of He gas was used to calibrate an X-ray spectrometer and at this point an emission line of He^+ was seen at a laser intensity four orders of magnitude lower than could be explained by invoking field ionisation alone. It was discovered that a small amount of Ar in the gas line ($\simeq 1\%$ from a previous calibration run) was clustering, being strongly heated by the laser pulse and exciting the surrounding He.

Several experiments have investigated absolute yields of X-rays from

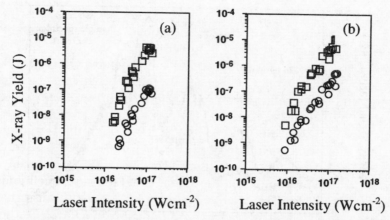

Fig. 7.19. The X-ray yield integrated over 4π from an Xe jet backed with 35 bars at temperatures of (a) 20 °C and (b) −85 °C. Squares are the yield measured by a diode with a 12-μm-Be filter (<1 keV X-rays); circles are the yield measured by a diode filtered with 15 μm Al (<3 keV X-rays), from [91].

cluster targets [91, 92], in part with the aim of characterising these sources for potential lithographic applications. There is no doubt that, although they are at present less efficient than solid targets by a factor of ten or so, laser-irradiated cluster media can produce copious amounts of X-rays in the 0.1–10 keV range. Chapter 5 describes work that used photon counting to estimate yields from a relatively low average density of 10^{14} atoms cm^{-3} but, at the significantly higher average densities produced directly under a pulsed gas jet, much more substantial yields of X-rays are produced. These are amenable to measurements with simple filtered PIN X-ray diodes as shown below. As with the data presented in chapter 5, there is a strong scaling of yield with laser intensity and the exact reason for this scaling is still being investigated.

Figure 7.19 shows time-integrated yield data from a Xe-cluster target recorded with magnetically shielded silicon PIN photodiodes, taken from [91]. A 12-μm-thick Be filter placed in front of the first diode blocked radiation with photon energies below 1 keV and a 15-μm-thick Al filter placed in front of the second diode removed most photons with energies below 3 keV.

The microjoule single-shot yields of X-rays seen in [91] and figure 7.19 also allow high-resolution spectroscopic studies using crystal spectrometers to be carried out. The low reflectivity of Bragg crystals in the kiloelectron-volt range means that these devices have low throughput. However, figure 7.20 shows that single-shot X-ray spectra from these cluster plasmas can be recorded quite easily. Experiments of this type also allow us to examine the strong scaling with pump-laser wavelength claimed to

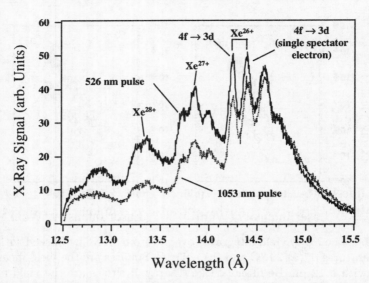

Fig. 7.20.   High-resolution single-shot X-ray spectra from a Xe-cluster plasma
taken using a Bragg-crystal spectrometer and a CCD detector. Data for 526-
nm (solid line) and 1053-nm (dotted line) laser pulses with identical gas-jet
conditions are shown, demonstrating the very weak scaling with wavelength.
Taken from [93].

have been observed in experiments by Kondo and co-workers [86], which
were interpreted using the inner-shell-ionisation mechanism of Rhodes
and co-workers [85, 87]. In the Kondo experiment various laser systems,
pulse durations and gas jets were used to attempt to compare absolute
X-ray yields with 800-nm and 249-nm irradiation. However, the very significant differences at these two pump-laser wavelengths reported in [86]
may well be due to experimental difficulties rather than a strong scaling with laser wavelength in the cluster-heating mechanism, a suggestion
that is also supported by the weak scalings with wavelength observed
in single-cluster-interaction experiments [94, 95]. Figure 7.20 shows a
very weak scaling of the yield of X-rays from Xe-cluster targets in the
12–15-Å regime for 1053- and 526-nm pulses, with all other experimental
parameters held constant, other than a slight shortening of the 526-nm
pulse during frequency doubling.

   The spectrum shown in figure 7.20 allows more detail to be seen than is
possible with the solid-state-semiconductor technique described in chapter
5 and some intriguing high-density plasma effects are now evident. Strong
emission from high charge states including Xe^{26+}, Xe^{27+} and Xe^{28+} (Ni-
like to Mn-like Xe) is clearly visible; however, at $\simeq 14.4$ Å the usual 4f–3d
transition in Xe^{26+} is accompanied by an additional shifted line from this
transition with a spectator electron present. Such an effect is typical of

high-density (near-solid) plasmas and the presence of strong satellite-line emission from the short-pulse irradiation of clusters has also been reported by Rhodes' group [87].

The ionisation kinetics of Xe have been investigated numerically to allow comparison with the data in figure 7.20 [92]. Calculations were carried out using GRASP [96], a multi-configurational Dirac–Fock code to find wavelengths and oscillator strengths of transitions in this region of the emission spectrum of ionised Xe. It was found that the strong line at 14.25 Å was a 4f–3d transition, which the code predicted to be the dominant transition in Ni-like Xe (Xe^{26+}). The feature at 14.4 Å is not a single line but rather a closely spaced group of Cu-like satellite lines. These satellite lines arise from 4f–3d transitions in ions that possess a single spectator electron in the 4s, 4p or 4d levels. The calculations revealed a high concentration of such transitions between 14.39 and 14.48 Å. The other broad features between 12.7 and 14.1 Å arise from unresolved 4f–3d transition arrays in the higher charge states of Xe, that is Co-like, Fe-like and Mn-like states.

7.11 Cluster jets as pulsed ion sources

Pulsed ion sources in which a laser ablates a solid surface and creates a transient plasma are now used routinely as a source of ions and electrons to seed particle accelerators. These 'photocathodes' are able to generate bursts of charged particles of extremely short duration, of order 10–100 ps [97]. The ability of cluster-interaction experiments to produce extremely high charge states [1, 2] has prompted an investigation of the ion fluxes from a bulk cluster target [98]. Ion time-of-flight measurements have been used by a number of groups to probe the dynamics of the energetic laser–cluster interaction [1–5]; however, these experiments typically use skimmer arrangements to ensure that the cluster beam in the interaction region is extremely dilute. Work has also been carried out [98] to quantify the ion flux from high-density cluster jets, both to provide insight into the dynamics of a bulk-cluster medium after the laser-heating phase (when the microplasmas from adjacent clusters begin to merge) and to investigate the utility of these sources for seeding heavy-ion accelerators.

Although energetic ions with energies well above 100 keV are seen to escape from bulk-cluster plasmas in work such as [47], more recent investigations [98] aimed at carefully characterising a number of laser-based ion sources for seeding accelerators found that cluster targets are not yet as efficient as solids for this type of application. While it is clear that laser irradiation of cluster targets can generate ions in very high charge states, there is a problem in having them escape from the bulk-cluster medium. This is to a great extent explained by the cluster-jet density profiles discussed in section 7.5. The spatial distribution of clusters from a pulsed gas jet does not have a sharp boundary, so ions created in the core of the jet must travel through a significant amount of material before they can escape. This problem could be solved by the use of suitable skimmer arrangements to scrape away the low-density wings of the jet, or by use of shaped supersonic nozzles, which result in a more tightly bounded gas flow.

7.12 Thermonuclear fusion in cluster plasmas

The very large energies of ions observed in cluster-interaction experiments and the high efficiency of coupling of energy to bulk-cluster media suggests a new method of driving thermonuclear fusion that can be achieved with small-scale (table-top) laser systems operating at high repetition rates. There are several schemes that might be amenable to such an investigation. The $D + T \rightarrow n + He^4$ reaction has by far the largest cross section in the energy range accessible by heating a cluster plasma with a short-pulse laser. However, the amount of tritium required to pressurise a cryogenically cooled pulsed-solenoid jet would represent a very serious radiological hazard and so to date only the D–D reaction has been investigated. With deuterium fuel, around half of the fusion events follow a $D + D \rightarrow n + He^3$ pathway and half a $D + D \rightarrow p + T^3$ pathway. Some small fraction ($\simeq 10^{-7}$) goes via a $He^4 + \gamma$ reaction. Of these three pathways, the neutron-producing reaction is the easiest and most definitive one to follow because neutrons can escape from a vacuum system and be detected and have their energy resolved using well-characterised scintillation techniques.

The first definitive observation of a cluster-based fusion reaction was made at the LLNL by Ditmire *et al.* [99] using a dense D_2-cluster target irradiated with a 35-fs, 150-mJ laser pulse from a Ti : sapphire CPA system that produced about 5×10^4 neutrons per shot. Previous work at Imperial College with a longer-pulse laser aided in design of the experiment and gas-jet system, but had been inconclusive insofar as the 2-ps pump laser used was less efficient at heating clusters of light atoms. Owing to the low value of the Hagena parameter k, deuterium needed to be cooled to 80 K or below to provide sufficiently large clusters. Neutrons were detected using plastic scintillators coupled to photomultipliers and their energy could be resolved using time-of-flight measurements. These showed that near mono-energetic 2.45-MeV neutrons were produced when a D_2-cluster target was used, but no signal was detected with H_2 clusters of comparable size (a useful null test because the p–p fusion cross section is many orders of magnitude smaller than the D–D cross section).

The extremely short-duration pulse of neutrons seen in these experi-

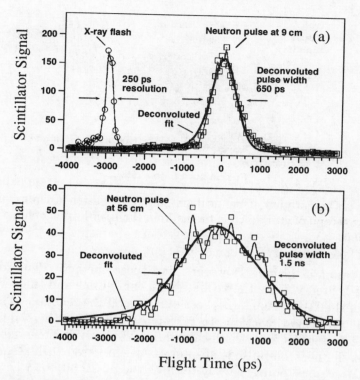

Fig. 7.21. (a) The measured neutron pulse width 9 cm from the cluster target. The deconvoluted pulse width is 650 ps. The detector's response (measured by producing hard X-rays from a solid target) is shown on the left-hand side. (b) The measured neutron pulse width 56 cm from the target. The pulse width is 1.5 ns. The detector's resolution was 1.1 ns.

ments and illustrated in figure 7.21 is of particular interest. Neutrons can be used in powerful methods for non-destructive probing of condensed matter and for structural analysis of X-ray-opaque objects. Pulsed neutron sources also have emerging applications in areas such as the remote analysis of chemical compounds through neutron-capture/gamma-ray-emission techniques. However, most neutron sources (Z pinches and spallation devices for example) have extremely long pulse durations (>400 ns) relative to the $\simeq$100 ps available from a cluster-fusion source, so new applications may well follow from this unique property. A useful issue to address here is that of why the neutron pulse from the cluster-fusion source is so short.

As a short laser pulse propagates into a dense D_2-cluster medium, local heating of clusters takes place in a cylindrical volume defined by the dimensions of the laser focus and the length of the cluster medium produced by the gas jet. Collisions between energetic deuterons from different

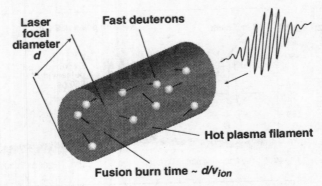

Fig. 7.22.   The duration of the neutron pulse in a cluster-fusion experiment is set by the escape of energetic deuterons from the laser-heated volume and is of order 100 ps.

clusters result in fusion. However, once these energetic deuterons leave the laser-heated volume they collide with cold atoms and the very strong scaling of the D–D-fusion cross section turns off the reaction. The 100-ps timescale for fusion reactions to take place is thus set by the escape of energetic deuterons from the focal volume. This short timescale and the fact that the neutron pulse is absolutely synchronised with respect to an intense laser pulse point the way towards the possibility of a new class of pump–probe experiments.

The single-shot yield seen in [99, 100] is quite modest compared with those of existing neutron sources (although the time-averaged flux is actually comparable to those of small-scale commercial systems). Applications such as long-term testing of components for future fusion reactors may well require higher yields and there are several ways in which the underlying laser–cluster interaction could be manipulated to achieve this. Increasing the size of cluster, for example, increases the average energy of deuterons up to a point in this experiment. However, this can introduce other problems associated with propagation of intense laser pulses in a highly absorbing cluster medium. Figure 7.23 shows that the yield of neutrons as a function of the size of cluster increases rapidly up to $\simeq 50$ Å. After that point the yield begins to fall off again and this is likely to be due to the increasing absorption of the laser pulse in the low-density wings of the jet. The strong n^2 density scaling of the D–D-fusion reaction limits the total yield of neutrons as a result. This could be avoided by 'machining' away clusters in the wings of the jet using a low-power UV laser as described in section 7.7. In addition, the energy of deuterons could be boosted by using several pulses [94] or an admixture of higher-Z ions [101] as described in chapter 6.

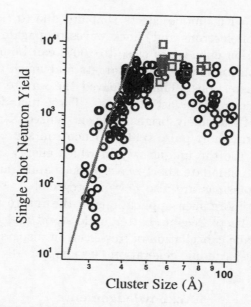

Fig. 7.23.   The neutron yield as a function of the size of cluster, from [100]. The yield first rises as larger clusters produce hotter ions and then falls as increasing absorption in the low-density wings of the jet prevents laser energy reaching the high density core.

7.13 Conclusions

In this chapter we have seen that cluster sources that can generate high average densities of atomic clusters (up to 5×10^{19} atoms cm^{-3}), with local densities close to solid, are quite easy to produce (though often troublesome to diagnose). Indeed, it seems likely that experiments in other areas of high-intensity laser–matter-interaction physics such as generation of harmonics in gas jets have inadvertently been producing and utilising extended cluster media for a good number of years. The failure of some of these experiments to give ever-increasing yields of harmonic radiation from targets such as Ar as the average density was increased could well have been the result of enhanced ionisation from clusters, coupled with additional free-electron dispersion that destroyed the phase-matching of the harmonics. Indeed, talking to people who work in this field, you hear a number of stories relating to 'mysterious' emission lines from these media, which were almost certainly due to ions in high charge states being produced in explosions of clusters.

Serendipitous observations aside, it is clear that extended cluster media embody the strong coupling to laser fields seen in interactions of isolated clusters, together with unique effects that are due to the way laser energy, superthermal electrons and shock waves propagate through such a medium. Absorption efficiencies close to 100% can be achieved in cluster media for both high- and low-Z targets and much of this energy is carried by the cluster ions. This has paved the way to a number of interesting applications in the last few years. These include demonstration of the principle of extremely bright, efficient pulsed X-ray sources and, more recently, ultra-short-pulse table-top neutron sources. More subtle experiments that use the unique way in which energy is deposited in a cluster medium to simulate shock-wave physics and transport of energy in astrophysical plasmas are also now in progress. It seems likely that we will see many more such applications in the future as the subtleties of single-cluster effects become better understood and applied to areas such as the dynamic cancellation of free-electron dispersion using cluster dispersion in high-harmonic generation (see chapter 6).

Acknowledgements

The work described in this chapter is the result of efforts by many people over a period of several years. We gratefully acknowledge their contributions to this chapter and the cluster-interaction field as a whole and would particularly like to thank J. Tisch for his many useful comments. We also thank the EPSRC and DoE for their funding of programmes carried out at Imperial College and the LLNL, respectively. The work at the LLNL was conducted under DoE contract W-7405-Eng-48.

References

[1] T. Ditmire, J. W. G. Tisch, E. Springate, M. B. Mason, N. Hay, J. P. Marangos, R. A. Smith and M. H R Hutchinson, *Nature* **386** 54 (1997).

[2] M. Lezius, S. Dobosz, D. Normand and M. Schmidt, *Phys. Rev. Lett.* **80** 261 (1998).

[3] S. Dobosz, M. Schmidt, M. Perdrix, P. Meynadier, O. Gobert, D. Normand, K. Ellert, T. Blenski, A. Y. Faenov, A. I. Magunov, T. A. Pikuz, I. Y. Skobelev and N. E. Andreev, *J. Exp. Theor. Phys.* **88** 1122 (1999).

[4] Y. L. Shao, T. Ditmire, J. W. G. Tisch, E. Springate, J. P. Marangos and M. H. R. Hutchinson, *Phys. Rev. Lett.* **77** 3343 (1996).

[5] T. Ditmire, E. Springate, J. W. G. Tisch, Y. L. Shao, M. B. Mason, N. Hay, J. P. Marangos and M. H. R. Hutchinson, *Phys. Rev. A* **57** 369 (1997).

[6] T. Ditmire, J. W. G. Tisch, E. Springate, M. B. Mason, N. Hay, J. P. Marangos and M. H. R. Hutchinson, *Phys. Rev. Lett.* **78** 2732 (1997).

[7] T. P. Martin, *Surf. Sci.* **156** 584 (1985).

[8] B. T. Heaton, *Phil. Trans. R. Soc.* **308** 95 (1982).

[9] O. F. Hagena, in *Molecular Beams and Low Density Gas Dynamics* Editior P. P. Wegener, Márcel Dekker Inc., New York, 1974.

[10] J. B. Anderson in *Molecular Beams and Low Density Gas Dynamics* Editior P. P. Wegener, Marcel Dekker Inc., New York, 1974.

[11] F. Spiegelmann, R. Poteau, B. Montag and P. G. Reinhard, *Phys. Lett. A* **242** 163 (1998).

[12] R. A. Smith, T. Ditmire and J. W. G. Tisch, *Rev. Sci. Instrum.* **69** 3798 (1998).

[13] O. F. Hagena, *Rev. Sci. Instrum.* **63** 2374 (1992).

[14] O. F. Hagena, *Rev. Sci. Instrum.* **62** 2038 (1991).

[15] J. P. Bucher, D. C. Douglass, P. Xia and L. A. Bloomfield, *Rev. Sci. Instrum.* **61** 2374 (1990).

[16] L. Bewig, U. Buck, C. Mehlmann and M. Winter, *Rev. Sci. Instrum.* **63** 3936 (1992).

[17] P. M. DevlinHill and H. T. Miles, *Rapid Commun. Mass Spectrosc.* **11** 541 (1997).

[18] D. B. Pedersen, J. M. Parnis and D. M. Rayner, *J. Chem. Phys.* **109** 551 (1998).

[19] U. Zimmermann, N. Malinowski, U. Näher, S. Frank and T. P. Martin, *Z. Phys. D* **31** 85 (1994).

[20] A. Behjat, G. J. Tallents and D. Neely, *J. Phys. D* **30** 2872 (1997).

[21] C. Altucci, T. Starczewski, E. Mevel, C. G. Wahlstrom, B. Carre and A. Lhuillier, *J. Opt. Soc. Am. B* **13** 148 (1996).

[22] C. A. Coverdale, C. B. Darrow, R. Jones, W. Sawyer, J. Crane, T. Ditmire, M. D. Perry and P. C. Filbert, *Rev. Sci. Instrum.* **66** 160 (1995).

[23] O. F. Hagena and W. Obert, *J. Chem. Phys.* **56** 1793 (1972).

[24] J. Wörmer, V. Guzielski, J. Stapelfeldt and T. Möller, *Chem. Phys. Lett.* **159** 321 (1989).

[25] O. F. Hagena, *Z. Phys. D* **4** 291 (1987).

[26] J. Arno and J. W. Bevan, *Jet Spectroscopy and Molecular Dynamics* Ed. J. M. Hollas and D. Phillips. Chapman and Hall, Blackie Academic and Professional, Glasgow, 1995.

[27] J. Farges, M. F. de Feraudy, B. Raoult and G. Torchet, *J. Chem. Phys.* **84** 3491 (1986).

[28] M. Lewerenz, B. Schilling and J. P. Toennies, *Chem. Phys. Lett.* **206** 381 (1993).

[29] A. L. Koch, *J. Theor. Biol.* **12** 276 (1966).

[30] T. Ditmire, *Phys. Rev. A* **57** R4094 (1998).

[31] J. D. Jackson, *Classical Electrodynamics* 2nd edn., J Wiley, New York, 1975.

[32] R. Klingelhöfer and M. H. Moser, *J. Appl. Phys.* **43** 4575 (1972).

[33] J. Gspann and K. Korting, *J. Chem. Phys.* **59** 1793 (1972).

[34] J. Gspann, *Surf. Sci.* **106** 219 (1981).

[35] M. Ehbrecht, B. Kohn, F. Huisken, M. A. Laguna and V. Paillard, *Phys. Rev. B* **56** 6958 (1997).

[36] B. von Issendorff and R. E. Palmer, *Rev. Sci. Instrum.* **70** 4497 (1999).

[37] D. C. Sperry, A. J. Midey, J. I. Lee, J. Qian and J. M. Farrar, *J. Chem. Phys.* **111** 8469 (1999).

[38] P. Lievens, P. Thoen, S. Bouckaert, W. Bouwen, F. Vanhoutte, H. Weidele, R. E. Silverans, A. Navarro-Vazquez and P. V. Schleyer, *J. Chem. Phys.* **110** 10316 (1999).

[39] R. Karnbach, M. Joppien, J. Stapelfeldt and J. Wormer, *Rev. Sci. Instrum.* **64** 2839 (1993).

[40] U. Buck and R. Krohne, *J. Chem. Phys.* **105** 5408 (1996).

[41] M. Lewerenz, B. Schilling and J. P. Toennies, *Chem. Phys. Lett.* **206** 381 (1993).

[42] C. Altucci, C. Beneduce, R. Bruzzese, C. D. Lisio, G.S. Sorrentino, T. Starczewski and F. Vigilante, *J. Phys. D* **29** 68 (1996).

[43] G. J. Pert, *Plas. Phys.* **16** 1051 (1974).

[44] G J Pert, *J. Fluid Mech.* **100** 257 (1980).

[45] M. Born and E. Wolf, *Principles of Optics* 6th edn., Pergamon, New York, 1980, pp 87–8.

[46] T. Ditmire and R. A. Smith, *Opt. Lett.* **23** 618 (1998).

[47] T. Ditmire, R. A. Smith, J. W. G. Tisch and M. H. R. Hutchinson, *Phys. Rev. Lett.* **78** 3121 (1997).

[48] M. Lezius, S. Dobosz, D. Normand and M. Schmidt, *J. Phys. B* **30** L251 (1997).

[49] H. M. Milchberg, R. R. Freeman, S. C. Davey and R. M. More, *Phys. Rev. Lett.* **61** 2364 (1988).

[50] M. M. Murnane, H. C. Kapteyn, M. D. Rosen and R. W. Falcone, *Science* **251** 531 (1991).

[51] T. Ditmire, E. T. Gumbrell, R. A. Smith, L. Mountford and M. H. R. Hutchinson, *Phys. Rev. Lett.* **77** 498 (1996).

[52] T. Ditmire, T. Donnelly, A. M. Rubenchik, R. W. Falcone and M. D. Perry, *Phys. Rev. A* **53** 3379 (1996).

[53] M. D. Perry, O. L. Landen, A. Szöke and E. M. Campbell, *Phys. Rev. A* **37** 747 (1988).

[54] J. W. G. Tisch, T. Ditmire, D. J. Fraser, N. Hay, M. B. Mason, E. Springate, J. P. Marangos and M. H. R. Hutchinson, *J. Phys. B* **30** L709 (1997).

[55] T. Ditmire, R. A. Smith and M. H. R. Hutchinson, *Opt. Lett.* **23** 322 (1998).

[56] Y. Nagata, K. Midorikawa, S. Kubodera, M. Obara, H. Tashiro and K. Toyoda, *Phys. Rev. Lett.* **71** 3774 (1993).

[57] A. Modena, Z. Najmudin, A. E. Dangor, C. E. Clayton, K. A. Marsh, C. Joshi, V. Malka, C. B. Darrow, C. Danson, D. Neely and F. N. Walsh, *Nature* **377** 606 (1995).

[58] R. Rankin, C. E. Capjack, N. H. Burnett and P. B. Corkum, *Opt. Lett.* **16** 835 (1991).

[59] P. Monot, T. Auguste, L. A. Lompré, G. Mainfray and C. Manus, *J. Opt. Soc. Am. B* **9** 1579 (1992).

[60] P. Monot, T. Auguste, P. Gibbon, F. Jakober, G. Mainfray, A. Dulieu, M. Louisjacquet, G. Malka and J. L. Miquel, *Phys. Rev. Lett.* **74** 2953 (1995).

[61] A. B. Borisov, A. V. Borovskiy, V. V. Korobkin, A. M. Prokhorov, O. B. Shiryaev, X. M. Shi, T. S. Luk, A. McPherson, J. C. Solem, K. Boyer and C. K. Rhodes, *Phys. Rev. Lett.* **68** 2309 (1992).

[62] C. G. Durfee III and H. M. Milchberg, *Phys. Rev. Lett.* **71** 2409 (1993).

[63] J. F. Luciani, P. Mora and J. Virmont, *Phys. Rev. Lett.* **51** 1664 (1983).

[64] P. A. Holstein, J. Delettrez, S. Skupsky and J. P. Matte, *J. Appl. Phys.* **60** 2296 (1986).

[65] A. Bendib, J. F. Luciani and J. P. Matte, *Phys. Fluids* **31** 711 (1988).

[66] P. Mora and J. F. Luciani, *Laser Particle Beams* **12** 387 (1994).

[67] P. Alaterre, C. Chenaispopovics, P. Audebert, J. P. Geindre and J. C. Gauthier, *Phys. Rev. A* **32** 324 (1985).

[68] T. J. Goldsack, J. D. Kilkenny, B. J. MacGowan, P. F. Cunningham, C. L. S. Lewis, M. H. Key and P. T. Rumsby, *Phys. Fluids* **25** 1634 (1982).

[69] B. T. V. Vu, A. Szöke and O. L. Landen, *Phys. Rev. Lett.* **72** 3823 (1994).

[70] T. Ditmire, E. T. Gumbrell, R. A. Smith, A. Djaoui and M. H. R. Hutchinson, *Phys. Rev. Lett.* **80** 720 (1998).

[71] M. Dunne, T. Afsharrad, J. Edwards, A. J. MacKinnon, S. M. Viana, O. Willi and G. Pert, *Phys. Rev. Lett.* **72** 1024 (1994).

[72] J. F. Luciani, P. Mora and J. Virmont, *Phys. Rev. Lett.* **51** 1664 (1983).

[73] P. Mora and J. F. Luciani, *Laser Particle Beams* **12** 387 (1994).

[74] L. Spitzer and R. Härm, *Phys. Rev.* **89** 977 (1953).

[75] W.L. Kruer, *The Physics of Laser Plasma Interactions* Addison-Wesley, Redwood City, 1988.

[76] A. Djaoui and A. A. Offenberger, *Phys. Rev. E* **50** 4961 (1994).

[77] A. Djaoui and S. J. Rose, *J. Phys. B* **25** 2745 (1992).

[78] T. Ditmire, K. Shigemori, B. A. Remington, K. Estabrook and R. A. Smith, *Astrophys. J. Suppl. Series* **127** 299 (2000).

[79] K. Shigemori, T. Ditmire, B. A. Remington, V. Yanovsky, D. Ryutov, K. G. Estabrook, M. J. Edwards and A. J. MacKinnon, *Astrophys. J.* **533** L159 (2000).

[80] C. F. McKee and B. T. Draine, *Science* **252** 397 (1991).

[81] R. Strickland and J. M. Blondin, *Astrophys. J.* **449** 727 (1995).

[82] D. Arnett, *Supernovae and Nucleosynthesis* Princeton University Press, Princeton, 1996.

[83] B. A. Remington, J. Kane, R. P. Drake, S. G. Glendinning, K. Estabrook, R. London, J. Castor, R. J. Wallace, D. Arnett, E. Liang, R. McCray, A. Rubenchik and B. Fryxell, *Phys. Plasmas* **4** 1994 (1997).

[84] Y. B. Zel'dovich and Y. P. Raizer, *Physics of Shock Waves and High-Temperature Hydrodynamic Phenomena* ed. W. D. Hayes and R. F. Probstein, Vol. 1, Academic Press, New York, 1966.

[85] A. McPherson, T. S. Luk, B. D. Thompson, A. B. Borisov, O. B. Shiryaev, X. Chen, K. Boyer and C. K. Rhodes, *Phys. Rev. Lett.* **72** 1810 (1994).

[86] K. Kondo, A. B. Borisov, C. Jordan, A. McPherson, W. A. Schroeder, K. Boyer and C. K. Rhodes, *J. Phys. B* **30** 2707 (1997).

[87] W. A. Schroeder, F. G. Omenetto, A. B. Borisov, J. W. Longworth, A. McPherson, C. Jordan, K. Boyer, K. Kondo and C. K. Rhodes, *J. Phys. B* **31** 5031 (1998).

[88] T. Ditmire, T. Donnelly, A. M. Rubenchik, R. W. Falcone and M. D. Perry, *Phys. Rev. A* **53** 3379 (1996).

[89] J. Larsson and A. Sjogren, *Rev. Sci. Instrum.* **70** 2253 (1999).

[90] B. E. Lemoff, G. Y. Yin, C. L. Gordon, C. P. J. Barty and S. E. Harris, *Phys. Rev. Lett.* **74** 1574 (1995).

[91] T. Ditmire, R. A. Smith, R. S. Marjoribanks, G. Kulcsar and M. H. R. Hutchinson, *Appl. Phys. Lett.* **71** 166 (1997).

[92] S. Dobosz, M. Lezius, M. Schmidt, P. Meynadier, M. Perdrix, D. Normand, J. P. Rozet and D. Vernhet, *Phys. Rev. A* **56** R2526 (1997).

[93] T. Ditmire, P. K. Patel, R. A. Smith, J. S. Wark, S. Rose, D. Milathianaki, R. S. Marjoribanks and M. H. R. Hutchinson, *J. Phys. B* **31** 2825 (1998).

[94] E. Springate, N. Hay, J. W. G. Tisch, M. B. Mason, T. Ditmire, M. H. R. Hutchinson and J. P. Marangos, *Phys. Rev. A* **61** 3201 (2000).

[95] R. A. Smith, J. W. G. Tisch, T. Ditmire, E. Springate, N. Hay, M. B. Mason, E. T. Gumbrell, A. J. Comley, L. C Mountford, J. P. Marangos and M. H. R. Hutchinson, *Physica Scripta* **T80A** 35 (1999).

[96] K. G. Dyall, I. P. Grant, C. T. Johnson, F. A. Parpia and E. P. Plummer, *Comput. Phys. Commun.* **55** 425 (1989).

[97] M. Uesaka, K. Kinoshita, T. Watanabe, T. Ueda, K. Yoshii, H. Harano, K. Nakajima, A. Ogata, F. Sakai, H. Kotaki, M. Kando, H. Dewa, S. Kondo, Y. Shibata, K. Ishi and M. Ikezawa, *Nucl. Instrum. Methods A* **410** 424 (1998).

[98] P. Fournier, H. Haseroth, H. Kugler, N. Lisi, R. Scrivens, F. V. Rodriguez, P. DiLazzaro, F. Flora, S. Duesterer, R. Sauerbrey, H. Schillinger, W. Theobald, L. Veisz, J. W. G. Tisch and R. A. Smith, *Rev. Sci. Instrum.* **71** 1405 (2000).

[99] T. Ditmire, J. Zweiback, V. P. Yanovsky, T. E. Cowan, G. Hays and K. B. Wharton, *Nature* **398** 489 (1999).

[100] J. Zweiback, R. A. Smith, T. E. Cowan, G. Hays, K. B. Wharton, V. P. Yanovsky and T. Ditmire, *Phys. Rev. Lett.* **84** 2634 (2000).

[101] J. W. G. Tisch, N. Hay, E. Springate, E. T. Gumbrell, M. H. R. Hutchinson and J. P. Marangos, *Phys. Rev. A* **60** 3076 (1999).

Index